COVER: **The English South Coast, English Channel and the Normandy assault area.** *(Dominic Stickland)*

First published in April 2013

A catalogue record for this book is available from the British Library

ISBN 978 0 85733 234 9

Library of Congress control no. 2012953860

Published by Haynes Publishing,
Sparkford, Yeovil, Somerset BA22 7JJ, UK
Tel: 01963 442030 Fax: 01963 440001
Int. tel: +44 1963 442030 Int. fax: +44 1963 440001
E-mail: sales@haynes.co.uk
Website: www.haynes.co.uk

Haynes North America Inc.,
861 Lawrence Drive, Newbury Park,
California 91320, USA

Printed in the USA by Odcombe Press LP,
1299 Bridgestone Parkway, La Vergne, TN 37086

Acknowledgements

My thanks go to the following individuals and institutions for permission to reproduce photographs from their collections: the Imperial War Museum, London; National Archives of Canada; the Royal Navy Submarine Museum, Gosport; iStock (Mikael Damkier and Brett Charlton); Tactical Air Reconnaissance Archive; Andy Thomas; US Coast Guard Collection; US Dept of Defense; US National Archives; US Navy.

The Fold3 website (www.Fold3.com) for making available online a vast selection of photographs from the US National Archives.

The Assault Glider Trust, Shrewsbury, in particular Richard Head, Martin Locke and Gary Wann, and for their kind permission to reproduce copyright photographs of the AGT's Horsa restoration.

Tim Beckett, for help with information and photographs about his father, Major Allan Harry Beckett.

Alf Sore, whose account of his experience on 1 Heavy Glider Maintenance Unit is reproduced with his permission.

My sincere thanks go to my friend Dr Simon Trew, Deputy Head of War Studies at RMA Sandhurst, for reading and commenting on my manuscript, and for his guidance on documentary sources. I am indebted to him for his great generosity of time and spirit.

Last but not least I would like to thank Sophie Blackman (project manager at Haynes), John Hardaker (copy editor), Rod Teasdale (page-builder), Dominic Stickland (artwork), Lee Parsons (cover design), Penny Housden (proof-reader) and Dean Rockett (indexer), for their help and support during the production of this book.

Every effort has been made to determine copyright ownership of the images that appear in this book. In some cases this has not been possible. The author apologises if any copyright has been accidentally infringed.

D-DAY

'Neptune', 'Overlord' and the Battle of Normandy

Operations Manual

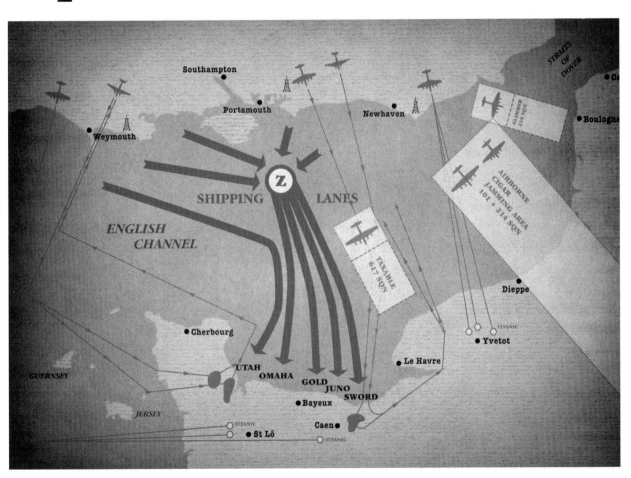

Insights into how science, technology and engineering
made the Normandy invasion possible

Jonathan Falconer

Contents

OPPOSITE GMC trucks of the US Army's 18th Regimental Combat Team (RCT) are loaded on to Landing Ship, Tank (LST) 134 on the English south coast at the beginning of June 1944 in preparation for Operation Overlord. This LST was one of 24 such vessels in Group 30 of LST Flotilla 12, Assault Force 'O' – destination Fox Green and Easy Red sectors of Omaha Beach on 6 June. *(US National Archives – USNA)*

Introduction

───(●)───

'This is the BBC Home Service – and here is a special bulletin read by John Snagge. D-Day has come. Early this morning the Allies began the assault on the north-western face of Hitler's European fortress. The first official news came just after half-past nine, when Supreme Headquarters of the Allied Expeditionary Force issued Communiqué Number One. This said: "Under the command of General Eisenhower, Allied naval forces, supported by strong air forces, began landing Allied armies this morning on the northern coast of France."'

OPPOSITE The Great Crusade: second-wave troops of the 9th Canadian Infantry Brigade, probably the Highland Light Infantry of Canada, disembark with bicycles from Landing Craft Infantry (Large) LCI(L) in the Juno Beach sector on to Nan White Beach at Bernières-sur-Mer, shortly before midday on 6 June. In the centre of the photograph a Small Box Girder (SBG) bridge (see page 108) has been laid over the sea wall by an AVRE Churchill, opening up an exit from the beach for wheeled vehicles.
(National Archives of Canada PA-131506)

RIGHT **Two of the principal Allied commanders on D-Day: 'Ike' – General Dwight D. Eisenhower, Supreme Allied Commander (right), and 'Monty' – General Bernard L. Montgomery, commander of Allied land forces in the invasion.** *(USNA)*

BELOW RIGHT
Support troops of the 3rd British Infantry Division assemble on Queen Red sector, Sword Beach, near La Brèche, Hermanville-sur-Mer, at 8:30am on 6 June, while under intermittent enemy mortar and shell fire. In the foreground and on the right, identified by the white bands around their helmets, are sappers of 84 Field Company Royal Engineers, part of No 5 Beach Group. Behind them, heavily laden medical orderlies of 8 Field Ambulance, Royal Army Medical Corps (some of whom are assisting the wounded) prepare to move off the beach. In the background can be seen men of the 1st Battalion, the Suffolk Regiment and Lord Lovat's No 4 Army Commando landing from Landing Craft Infantry (Small) LCI(S) of Naval Force S. *(IWM B5114)*

The world awoke on 6 June 1944 to the momentous news of the Allied invasion of France – the greatest amphibious assault in history. Operation Overlord had begun and on D-Day alone 156,000 troops (including some 20,000 airborne) were landed in Normandy, supported by 6,939 vessels (including the build-up force due to land troops on D+1 and D+2, plus reserves) and 14,674 sorties by Allied aircraft. By the end of 11 June (D+5), 326,547 troops, 54,186 vehicles and 104,428 tons of supplies had been landed over the beaches and by air.

To launch a frontal attack from the sea was a risky business and its success depended on meticulous planning and bold implementation. Before the first troops stepped ashore, audacious landings by paratroopers and glider-borne infantry had seized key positions on the flanks of the invasion area in what was to be the first – and the last – mass parachute drop to be attempted in darkness.

LEFT US 9th Air Force C-47 Skytrains, towing Horsa gliders, overfly Utah Beach on the evening of D-Day. *(USNA)*

Preparations for Overlord and Neptune (the naval component of the invasion) had indeed been meticulous, involving thousands of people on both sides of the Atlantic and with almost nothing left to chance. The Germans knew that an invasion was coming, but until the first troops hit the beaches of Normandy they did not know where. Even then they were uncertain as to whether or not this was a feint for main landings elsewhere.

D-Day was also important for a number of reasons: it saw paratroopers and glider-borne infantry used in large numbers; it was the first occasion in the European Theatre of Operations where purpose-built amphibious landing ships and craft were used in significant numbers; and the first in which massive tactical air support had been used before, during and after the landings.

It was also the first large-scale invasion in history where science, technology and innovation had played a truly major part – from tank-carrying gliders, swimming tanks, specialist engineer vehicles, to radio navigation aids that ensured coastal minefields were swept and that landing craft arrived on the right beaches, and floating command and control centres for the direction of Allied combat aircraft overhead; and from providing Allied forces with supplies and fuel, to the building of advanced airfields close to the front line.

D-Day – The background

The operation to land forces on the northern coast of France on what became known as D-Day, was a combined British and American undertaking. In the words of the Allied planners, the object of Operations 'Neptune' (the naval element) and 'Overlord' (the land forces element) was 'to secure a lodgement on the Continent from which further offensive operations can be developed'.

The main features of the 'Overlord' plan were the dropping of three airborne divisions on the eastern and western flanks of the invasion beaches before the main assault went in from the sea by five infantry divisions, Commandos and US Rangers, which were delivered by ships and landing craft along a 50-mile front between Ouistreham and Varreville. They would be consolidated by follow-up forces on the second tide of D-Day, then by the remainder of the follow-up forces on 7 June. Thereafter, the plan was to increase the strength of the land forces at a rate of one and one-third divisions each day. Once a firm lodgement had been established the objects were to capture the port of Cherbourg and then sweep south to occupy the Brittany ports within 35 to 40 days. Ultimately, the planners were looking at the annihilation of the German armies in the West, the capture of Paris and the liberation of southern France.

ABOVE An American
soldier runs for cover
as an M5 Stuart
light tank engages
a German position
in the Normandy
countryside. (USNA)

Festung Europa

But cracking open the walls of *Festung Europa*
(Fortress Europe) as the Germans called it, was
going to be a tough proposition for any invader.
The Nazis had invested massive resources in
building extensive and powerful fortifications
along the coastline of northern Europe (the
Atlantic Wall), with the aim of frustrating an
assault for long enough to enable the defenders
to mobilise their forces, repulse the raiders and
push them back into the sea; they had sown
defensive minefields off the French coast and

constructed a chain of radars to give early
warning of the approach of any invasion force.
The Luftwaffe, although seriously weakened by
the Allies' strategic air offensive, was still a force
to be reckoned with and the Kriegsmarine's
U-boat force, operating out of bases on the
Biscay coast, could wreak havoc with the
invasion convoys in the Channel.

About this book

'**D**-Day', as used in the main title of this
book, has become a shorthand for the
entire Normandy campaign that covered the
period from 6 June to late August 1944. This
Haynes D-Day Manual contains some chapters
that relate to D-Day proper – that is, 6 June
– and others that cover some aspects of the
wider campaign in Normandy.

There are two subjects in particular that lend
themselves to the 'see how it's done' Haynes
Manual treatment, which is why they have
been included in this book and are covered
in greater detail relative to some of the other
chapters. These are the Mulberry harbours and
the Advanced Landing Grounds. Neither had
anything to do with D-Day itself, but the former
(Mulberry) gave the Allies confidence to mount the

RIGHT Dishevelled
German troops are
surrounded at a
Normandy farm by
American troops
carrying M1 carbine
rifles. (USNA)

Geography of the assault

The D-Day assault area was defined as being bounded on the north by the parallel of latitude 49° 40' N, and on the west, south and east by the shores of the Bay of the Seine. This area was divided into two Task Force Areas, the boundary between them running from the root of the Port-en-Bessin western breakwater in an 025° direction to the meridian of longitude 0° 40' W, and thence along this meridian to latitude 49° 40' N.

The 1st US Army commanded by Lt-Gen Omar M. Bradley was to operate in the Western Task Force area, of which Rear-Admiral A.G. Kirk US Navy was the Naval Commander; and the 2nd British Army, commanded by Lt-Gen Miles Dempsey, in the Eastern Task Force area, with Rear-Admiral Sir Philip Vian RN as Naval Commander.

The Western Task Force area was divided into two assault force areas – the 'Utah' area covering the east coast of the Cotentin Peninsula to the River Vire; and the 'Omaha' area from there to the British area. Two Naval Assault Forces, 'U' and 'O' respectively, were responsible for all naval operations in these areas.

Airborne troops of the US 82nd and 101st Airborne Divisions were landed and dropped in the Cotentin Peninsula.

The Eastern Task Force area was divided into three assault force areas – 'Gold', from Port-en-Bessin to Ver-sur-Mer; 'Juno' from there to west of Langrune; and 'Sword' from there to Ouistreham – served by Naval Assault Forces 'G', 'J' and 'S' respectively.

Airborne troops of the British 6th Airborne Division (less 5th Parachute Brigade) were landed and dropped in the area east of Caen and astride the River Orne and Caen Canal.

BELOW The Normandy assault area, June 1944.
(Dominic Stickland)

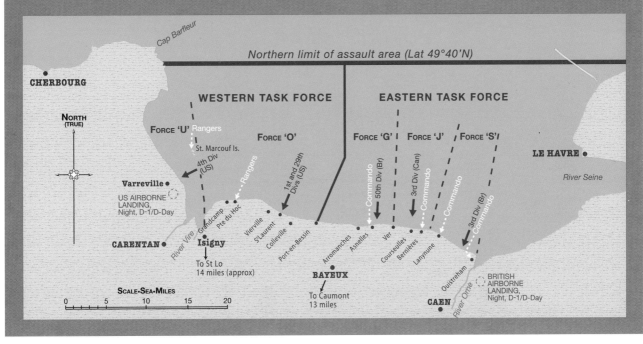

initial assault between, rather than against, major Channel ports; while the latter (Advanced Landing Grounds or forward airfields) allowed them to discount the Pas de Calais for the assault (where there were many more existing airfields) than might otherwise have been the case.

In a book of this size it is not realistic to attempt to cover the multitude of different systems and equipment used in ground, air and naval operations throughout the whole of the Normandy campaign, so this has been avoided. Instead, this manual offers insights into the design, construction and use of some of the innovative machines, structures and systems that were used on D-Day and after, revealing how they contributed to the success of Operations Overlord and Neptune, which paved the way for victory in Europe.

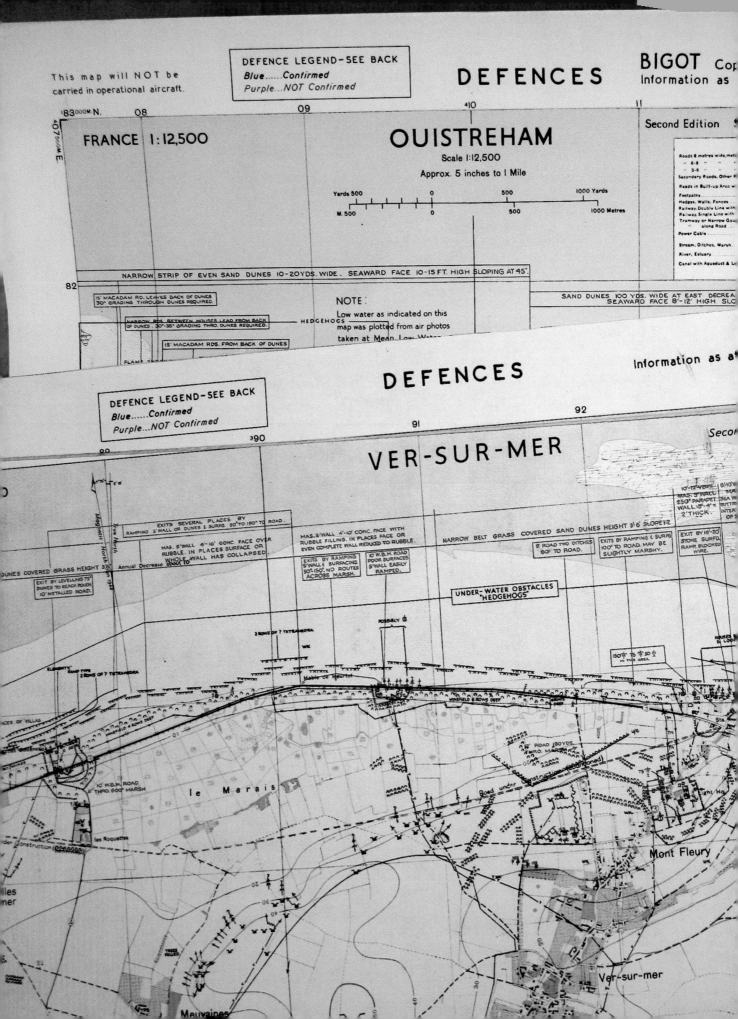

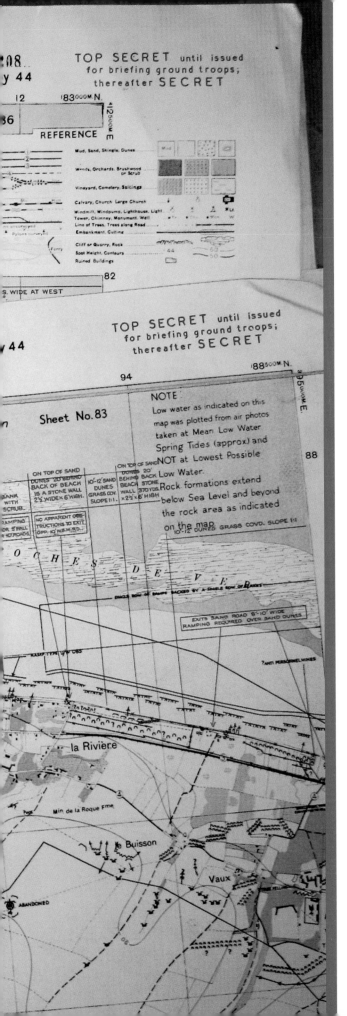

Chapter One

Men in white coats

The appliance of science

Science and technology, laced with human courage, made certain that D-Day would happen. Aerial photography, geology and hydrography, radio countermeasures and electronics all played their part in the choice of suitable landing beaches; ensured that the sea channels to the Normandy coast were cleansed of mines; and kept the enemy guessing as to where and when the invasion would come.

OPPOSITE It took the combined efforts of RAF and USAAF photoreconnaissance pilots, COPP clandestine beach survey teams and Royal Engineers air-photo mapping specialists to produce the detailed maps that were needed for the invasion. (*Author*)

BELOW This high-
altitude vertical
photograph of the
shoreline at Vierville
was taken on 24 May
1944 by an RAF
photoreconnaissance
(PR) Spitfire of 140
Squadron using an F52
camera fitted with a
36in lens. It shows the
western end of Omaha
Beach, codenamed
Dog Green, where two
weeks later American
troops were to suffer
huge casualties in
the landings. (Tactical
Air Reconnaissance
Archive)

Making charts and measuring beach gradients

Information on the gradients of possible landing beaches in northern France was vital to the D-Day planners. Trials conducted in 1942 concluded that beach gradients could be measured if a shoreline was covered with six lines of vertical aerial photographs, with one line run at (or close to) low-water springs, one line at (or again close to) high-water springs, and the remaining four lines evenly distributed between these two extremes. The gradient could be calculated by measuring the distance to each of the water lines and using the height of tide at the times the photographs were taken to arrive at a definitive figure.

To get the best possible image contrast between land and sea (to determine the water line) the photographs needed to be taken at very precise times – during the first two or last two hours of useable daylight. Wind speed was also a critical factor. It could not be more than 20 knots, as higher winds drove waves up the beach to give a false water line height. And to round off these very exacting demands, there had to be no clouds to obscure the location. As if this was not hard enough, there was the added difficulty of finding the conditions when the height of the tide was just right, and many photoreconnaissance (PR) sorties had to be scrubbed because of poor weather, or offensive sweeps that were being flown in the area of the sortie.

The RAF's 140 Squadron was tasked with gathering the photographs, which it began doing in May 1942 using its PR Spitfires (joined in March 1944 by Mosquitoes) flying initially from RAF Benson in Oxfordshire and later from the airfields at Mount Farm, Hartford Bridge (Blackbushe) and Northolt. As soon as a line of photographs had been obtained and processed, a small group of Royal Engineer surveyors from 1 Air Survey Liaison Section (also based at Blackbushe airfield) would create a photo mosaic from the pictures and then measure the distances to the water line. At the same time, the Tidal Branch of the Hydrographic Department at Taunton would supply the predicted tidal heights for the times the photographs had been taken. Once all six lines were complete for a given beach, the water lines from the five higher-tide mosaics would be transferred to the low-water mosaic. This, together with the measured distances and predicted tide heights, was duly passed to the Hydrographic Department, which would create the final gradient chart.

As D-Day approached events began to move rapidly, and from February 1944 the pressure increased on 140 Squadron's aircrew and the RE surveyors. PR coverage of enemy coastal defences, beaches and ports became more frequent, the seabed off the Calvados coast of Normandy was photographed and vast mosaics of Normandy and the Pas de Calais were completed. All this was in parallel to RAF Benson's other duties photographing bomb damage, railways, marshalling yards, bridges, airfields and V1 flying-bomb sites.

The 2½-mile-wide coastal strip that stretched from Blankenberghe in Belgium to

Avranches in Lower Normandy, which at first had been covered by the RAF's PR aircraft once every three months, was by this time being photographed once every three days by its Spitfires and Mosquitoes. Certain beaches were also being overflown at 6,000ft by PR aircraft equipped with ciné cameras to obtain large-scale imagery of the beach obstacles erected by the Germans to obstruct amphibious landings.

Despite the difficulties in getting photographs of the beaches at the right times and under the required conditions, 140 Squadron completed their task in May 1944, by which time some 200 beaches had been covered.

Dicing sorties

The requirement remained for more detailed photographs of the German beach defences along the Normandy coast, which was a task that fell to the USAAF's 10th Photo Reconnaissance Group (10th PRG). Using an extremely low-level technique developed by the USAAF's 3rd PRG in Italy, the first in a series of so-called 'dicing' missions over Normandy was flown on 6 May by Lt Albert Lanker of the 31st Photo Squadron (31st PS) in a single-seat Lockheed F-5E Lightning (the photoreconnaissance version of the P-38

ABOVE USAAF ground crewman Sgt Thomas Bowen opens the cowling that houses one of two sideways-facing K-17 cameras on 33rd PRS Lockheed F-5 'Super Snooper'. Note the forward-facing camera and the different mission symbols that include 18 dice, signifying that this aircraft had flown as many dicing sorties. *(US National Archives/Fold3)*

BELOW Control boxes inside the cockpit of an F-5, which were used by the pilot to operate the three recce cameras. Note the pilot's finger on the 'runaway control' switch. *(US National Archives/Fold3)*

ABOVE Detailed information on the Normandy beach defences was obtained from 'dicing' shots like this. Element 'C', steel gate-like boat barricades, are prominent in this picture (they failed to stop the landings). This photograph was taken on the first dicing mission flown by the USAAF's 10th Photo Reconnaissance Group on 6 May 1944, by 1st Lt Albert Lanker. *(US National Archives/ Fold3)*

Lightning fighter). The 10th PRG (comprising the 30th, 31st and 34th Photo Reconnaissance Squadrons) went on to fly 35 'dicing' missions from their base at Chalgrove in Oxfordshire between 6 and 20 May 1944.

The 'dicing' camera configuration for an F-5 consisted of a forward-facing K-17 camera fitted with a 12in focal length lens, tilted downwards at a 10° angle; and two K-17 cameras with 6in focal length lenses, one on each side of the forward fuselage behind the nose, aimed slightly forward from right angles to the aircraft's line of flight. This camera combination gave an uninterrupted coverage of more than 180°.

Taking off from Chalgrove, an F-5 pilot would fly at very low level over the Channel to make landfall over his respective area of coastline before turning to make the camera-run. Generally the beach was centred, with the land to the pilot's left and the Channel to his right. He would then set the cameras at 'runaway speed' (the shutter to fire every six seconds), push open the throttles and head out along the beach at a height of 25ft and a speed of about 375mph. With all three cameras firing quickly and simultaneously, with overlapping coverage, a broad cover of the target could be obtained in one pass.

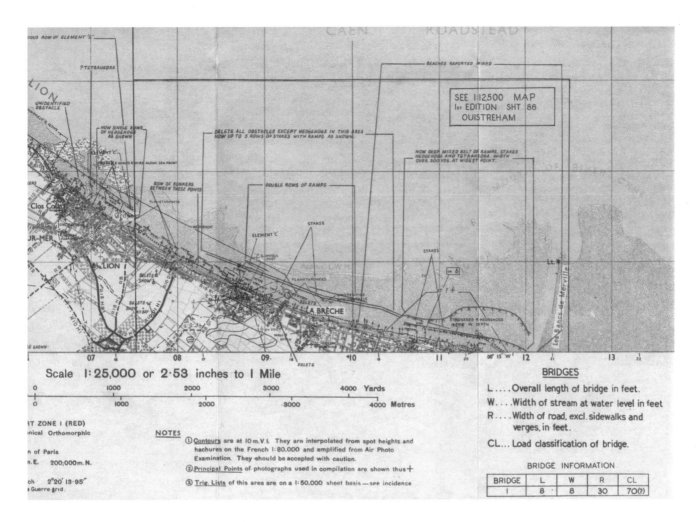

Scale 1:25,000 or 2·53 inches to I Mile

| | 0 | 1000 | 2000 | 3000 | 4000 Yards |
| | 0 | 1000 | 2000 | 3000 | 4000 Metres |

BRIDGES

L .... Overall length of bridge in feet.
W .... Width of stream at water level in feet
R .... Width of road, excl. sidewalks and verges, in feet.
CL ... Load classification of bridge.

BRIDGE INFORMATION

BRIDGE	L	W	R	CL
I	8 ·	8	30	7O(?)

NOTES

① Contours are at 10 m.V.I. They are interpolated from spot heights and hachures on the French 1:80,000 and amplified from Air Photo Examination. They should be accepted with caution.

② Principal Points of photographs used in compilation are shown thus +

③ Trig. Lists of this area are on a 1:50,000 sheet basis — see incidence

Flying an unarmed aircraft at such low level in broad daylight was 'dicey', as the mission name implied. On the second dicing mission to be dispatched F-5 pilot Lt Fred Hayes of the 31st PS was lost without trace on 7 May, but overall the missions were an unqualified success. The coverage they brought back showed Hedgehogs, Belgian Gates and the other deadly anti-invasion paraphernalia planted on the sands by the Germans, all in fine enough detail to give Allied commanders a clear idea of what the assault force would be up against when the time came.

Air photography for mapping

Photographic coverage of the Normandy coastline was actually part of a much larger mapping project. In addition to investigating beach gradients, reconnaissance aircraft of 140 Squadron were engaged in air photography for mapping purposes. When planning for D-Day began in 1942 there was

ABOVE This is a section of map from 'France 1:25,000 sheet 40/18 SW St Aubin, stop press edition 20 May 1944'. It is one of the many 1:25,000-scale maps produced from information supplied by RAF photoreconnaissance sorties over Normandy – one of the so-called 'Benson' block. It shows part of Sword Beach. The base map is a Geographical Section General Staff (GSGS) 4347 series sheet, with overprint detail showing the latest intelligence gleaned from air photography – for example, onshore and offshore defences. These maps, which covered the whole of the invasion coast, were produced under extreme secrecy and classified 'Top Secret' with the caveat 'Bigot', which meant they were only available to a small number of very senior staff who knew the full invasion plan. The blue overprint detail shows information from intelligence sources and air photography up to May 1944; the orange overprint updates this information to 19 May. *(Author's collection)*

insufficient mapping available to assist in the production of 1:25,000 and 1:12,500 sheets, so a major survey flying programme was instigated to acquire the survey-standard imagery necessary to produce *ab initio* mapping by air survey methods. This project was codenamed 'Benson' after the RAF photoreconnaissance airfield. For this task, six General Survey Sections of the Royal Engineers were raised and trained in air-photo mapping, in addition to the specialist 1 Air Survey Liaison Section working alongside the RAF.

The 'Benson' sheets provided the base mapping for a range of different overprints, some showing minute intelligence detail of German anti-landing craft obstacles, bunkers and gun emplacements. Revised editions of these overprints were being produced almost up to the eve of D-Day. However, a problem that remained temporarily unresolved by air reconnaissance was the identification for mapping purposes of detailed topographical features, specifically land contours and spot heights.

The geology of the beachhead

For more than a year before D-Day, British geologists had been involved in the planning of Overlord. During this time their two principal tasks were the study of likely invasion beaches and the soil conditions of possible airfield sites inland. Other geological activities included the study of cliffs in the invasion area, the provision of information about the foundations of enemy defences to enable their effective bombing by the RAF and USAAF, the preparation of water intelligence maps, the provision of information on sources of road metalling, sand and gravel, the submarine geology of the ports in Normandy, the detailed study of certain rivers with a view to assault crossings, and the nature of the sea floor beneath the Channel (for the Pluto petrol pipeline).

Extract from 140 Squadron Operations Record Book for May 1944

Date: 10 May
Aircraft type and number: Mosquito IX, MM249
Crew: F/L W. Shearman, F/O J. Bird
Duty: Beaches, Port en Bessin. Sp. 36in cameras [split pair F52 36in cameras]
Time up: 07.45hrs. Time down: 09.45hrs
Details of sortie: Required photographs obtained. Aircraft was intercepted over Cherbourg by two Spitfire Mk XIVs, which later identified the Mosquito before opening fire [sic].

Aircraft type and number: Mosquito XVI, MM302
Crew: F/O D.R. Thompson, F/Sgt Richardson
Duty: Beach, Boulogne. Sp. 36in cameras
Time up: 07.50hrs. Time down: 09.55hrs
Details of sortie: Photographs obtained after diving through cloud down to 12,500ft.

Date: 12 May
Aircraft type and number: Mosquito IX, MM249
Crew: F/O J.B. Reynolds, F/L C.G. Chadwick
Duty: Port en Bessin–Ouistreham. Sp. 36in cameras
Time up: 09.00hrs. Time down: 11.05hrs
Details of sortie: Photography of required beach obtained without incident. Aircraft diverted to Tangmere owing to bad visibility at base.

Note: The Mosquito PR IX carried five cameras: a double pair of split-pair vertical cameras (on the sorties above they were the F52 high-altitude day reconnaissance camera fitted with a 36in lens) and one oblique camera. The split pairs gave stereoscopic coverage of a target and its surrounding area and allowed for greater coverage of the ground on a single reconnaissance run.

Major General Sir J.D. Inglis, Chief Engineer of the 21st Army Group, recorded:

We had, fortunately, long appreciated the importance of geology in modern war, and at that time had the services of Professor King [Prof W.B.R. King, Professor of Geology at the University of London and from 1943 at Cambridge] who had been with us since the beginning of the war... Amongst other extremely valuable advice Professor King pointed out that between Caen and Bayeux there was a patch of country that was not only gently undulating but also possessed a topsoil that was particularly suitable for airfields because of its excellent drainage qualities. This was, in fact, one of the main factors that led to the selection of the beaches eventually used.

Once the Calvados coast had been chosen as the location for invasion, the character of the beaches and coastal cliffs as well as the soils and topography immediately inland, where the first phase of the battle would be fought, became the geologists' particular concern. They knew how important it was for vehicles to follow firm routes over the actual landing beaches.

A hard lesson had been learned on the beaches at Dieppe in 1942 where the Churchill tanks, in their baptism of fire, found it difficult to gain traction on the loose shingle composed of chert. This compact rock, consisting mainly of microcrystalline quartz, was known to be a substance that was considerably harder for tracked vehicles to negotiate than shingle. The churning of tank tracks trying to find traction

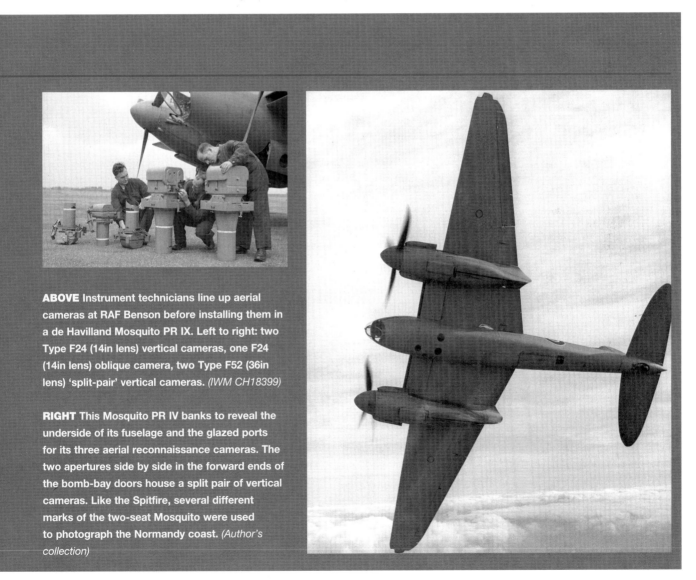

ABOVE Instrument technicians line up aerial cameras at RAF Benson before installing them in a de Havilland Mosquito PR IX. Left to right: two Type F24 (14in lens) vertical cameras, one F24 (14in lens) oblique camera, two Type F52 (36in lens) 'split-pair' vertical cameras. *(IWM CH18399)*

RIGHT This Mosquito PR IV banks to reveal the underside of its fuselage and the glazed ports for its three aerial reconnaissance cameras. The two apertures side by side in the forward ends of the bomb-bay doors house a split pair of vertical cameras. Like the Spitfire, several different marks of the two-seat Mosquito were used to photograph the Normandy coast. *(Author's collection)*

ABOVE Memories of the Dieppe fiasco on 19 August 1942 were invariably uppermost in the minds of Overlord's planners. It was a disaster they were determined to learn from. Here, burnt-out tanks and landing craft lie strewn across the beach at Dieppe after the Allied withdrawal. Of the 24 landing craft that took part, 10 managed to land a total of 24 tanks, all of which were lost. Canada's Churchill-equipped Calgary Regiment (14th Canadian Tank Regiment) was chosen to provide tank support for the infantry (also Canadian) destined to go ashore. The tank in the foreground is a Churchill III, 'Calgary', T68559, of C Squadron, HQ Fighting Troop, commanded by Lt B.G. Douglas. It was abandoned when its left track was broken by enemy shellfire. The landing craft is LCT 5, which was hit by mortars and its bridge destroyed, killing the captain and crew. *(Author's collection)*

ABOVE Captain Nigel Clogstoun-Willmott (1910–92) was the original inspiration behind the Combined Operations Pilotage Parties, or 'COPPists' as they were known. As a lieutenant-commander he carried out two surveys of the Normandy coast in 1944, one by landing craft (for which he was mentioned in dispatches) and the second by midget submarine, for which he was awarded a bar to his DSC. Much to his regret, illness prevented him from taking part in the actual landings. *(Copyright unknown)*

ABOVE To avoid the difficulties faced by tanks on the beaches at Dieppe, it was deemed vital to determine the geology of the proposed landing beaches, their gradient and load-bearing characteristics, and the existence of satisfactory routes inland. This is Gold Beach on 7 June, with soft-skinned transport crossing its firm sands at low tide. *(IWM B5125)*

dug deep trenches in the beach, forcing stones up between the tanks' drive wheels and tracks, causing them to fail. Some of the tanks became stranded, making them sitting targets for the German gunners.

To avoid a costly repeat of Dieppe, the Normandy beaches were analysed in detail, not only to determine their appearance, gradient and exits, but also the distribution of the patchy peat, clay, sand and shingle that formed the surface. Maps to 1:5,000 scale were prepared from existing topographical and geological records, aerial photographic reconnaissance and covert ground reconnaissance.

To confirm these details the beaches were covertly sampled on a moonless New Year's Eve in 1943 by a Combined Operations Pilotage

ABOVE The Combined Operations shoulder badge was worn by men of the Combined Operations Pilotage Parties, who were also known as 'COPPists'. *(Author's collection)*

Party (COPP) comprising Major Logan Scott-Bowden and Sgt Bruce Ogden-Smith of the Royal Engineers. The two-man team boarded a motor gunboat at Gosport, which took them to within a few miles of their destination, Luc-sur-Mer (on what was to become the British 'Gold' Beach) where the men changed into bulky rubber swimsuits and strapped on their heavy bandoliers, backpacks, measuring equipment and weapons. Then they transferred to a small inshore craft that took them to within a quarter of a mile of the beach from where they swam ashore. There they quartered the target beach, taking measurements and core samples with metal augers, storing them in special containers for analysis back in London by the British Geological Survey.

Scott-Bowden remembers:

> As we swam back through heavy surf towards our rendezvous point I thought my companion was in trouble when I heard him shouting. But when I turned to help he only wished me 'A Happy New Year'. I told him to, 'swim you b*****, or we'll land back on the beach'. However, I wished him 'A Happy New Year' in return and we used our infrared torches to signal to our

support boat, and then rendezvoused with the MGB. Weather conditions were against us on our return trip, and we had to head for Newhaven because we needed to make landfall before dawn and the possibility of our discovery by the enemy.

BELOW Major Scott-Bowden and Sgt Ogden-Smith were taken across the Channel in a Royal Navy motor gunboat, similar to this C-type Fairmile MGB, to carry out their clandestine beach sampling sortie. *(Author's collection)*

Operation Postage Able

When the Americans got to hear of this daring mission, they asked for a COPP party to survey their own landing sites during the next moonless period a few weeks later. On 16 January, X-craft midget submarine *X20* was towed from Portsmouth by a Royal Navy trawler to within a few miles of the French coast in Operation Postage Able. *X20* was commanded by Lt Ken Hudspeth DSC & Bar, RANVR (a veteran of the X-craft attack on the German battleship *Tirpitz* in 1943), and Sub-Lt Bruce Enzer RNVR, with the COPP team comprising Lt-Cdr Nigel Clogstoun-Willmott DSO, DSC, RN, and again Major Logan Scott-Bowden DSO, MC and Sergeant Bruce Ogden-Smith DCM, MM.

Scott-Bowden recalls:

We then spent four days on the seabed, and three nights surveying the beaches to the west of Port-en-Bessin and the 'Omaha' beach area. When we first approached the target area we discovered our way was partially blocked by a French fishing fleet, complete with enemy guards. Our Australian skipper Ken Hudspeth said we could work our way in under the nets, and as we threaded our way through, we raised the periscope. I was a little surprised to find myself staring into the face of a German soldier, perched close by up on the stern of the last fishing boat, thoughtfully puffing away on a pipe! We downed periscope pretty smartly I can tell you!

During the hours of daylight the midget submarine lay submerged at periscope depth while its crew conducted a reconnaissance of the shoreline and took bottom soundings using an echo sounder. Each night *X20* surfaced and moved in closer to the shore to allow Scott-Bowden and Ogden-Smith to swim some 400yd to the beach to carry out their geological survey work at Vierville, Les Moulins and Colleville (the American 'Omaha' beach sector).

Wearing a wet suit, each man was weighed down with a shingle bag, brandy flask, sounding lead, underwater writing pad and pencil, compass, beach gradient reel and stake, .45 revolver, trowel, auger, torch and bandolier. The soil samples they collected from the beach were placed inside condoms to protect them from water on the return swim.

On the third night Scott-Bowden and Ogden-Smith were to have gone ashore further up the coast around the Orne estuary, but

LEFT Lieutenant Ken Hudspeth, RANVR, commanded X-craft *X20* on D-Day. He was awarded the Distinguished Service Cross (DSC) and two bars in the space of 11 months in 1944, making him one of Australia's most decorated naval officers of the Second World War. The citation for the second bar to his DSC was for 'gallantry, skill, determination and undaunted devotion to duty'. When the war ended he returned to a career in teaching. He died in 2000 aged 82. *(Author's collection)*

by this time fatigue had set in. The five men inside *X20* had been living on little more than water and Benzedrine tablets (amphetamines – which were used as a stimulant and appetite suppressant) which, combined with the worsening weather, caused Hudspeth to cut short the operation and return to HMS *Dolphin* on 21 January. This operation gathered a great deal of information that proved invaluable in the planning for D-Day, for which Hudspeth was rewarded with a bar to his DSC.

Operation Gambit

Ahead of D-Day, Ken Hudspeth (once again operating in *X20*, but this time accompanied by Lt George Honour RNVR in *X23*) was ordered back to Normandy to assist off Juno and Sword beaches in guiding in the main assault fleet. They arrived in position on 4 June, but when bad weather in the Channel caused postponement of the landings they were compelled to remain submerged and in position until 4:30am on 6 June (D-Day). They surfaced and set up their navigational aids – an 18ft telescopic mast with a green light shining seaward and visible up to five miles away, a radio beacon and an echo-sounder tapping out a message for the minesweepers approaching these beaches.

LEFT Lieutenant George Honour DSC, RNVR (1918–2002), a Bristolian, joined the RNVR in 1940. He commanded *X23* on D-Day. *(Royal Navy Submarine Museum)*

ABOVE The first craft off the shores of Normandy on D-Day were two Royal Navy midget submarines – X-craft *X20* and *X23*. This rare photograph of *X23* on D-Day shows Lt George Honour and Sub-Lt H.J. Hodges RNVR on the casing. *(Royal Navy Submarine Museum)*

Electronic countermeasures

The USAAF's offensive against the Luftwaffe's fighter force in north-west Europe during 'Big Week' (20–26 February 1944) saw the fortunes of the German Air Force take a downward turn. Even so, through the winter of 1943–44 RAF Bomber Command had suffered grievous losses in the hard-fought Battle of Berlin, when 1,117 Lancasters and Halifaxes fell to the guns of Luftwaffe fighters and flak batteries. The infamous Nuremberg raid of 30–31 March saw 95 bombers out of a force of 795 downed by night-fighters (the biggest Bomber Command loss of the war); and the raid on the German military camp at Mailly-le-Camp on 3–4 May saw 42 Lancasters shot down out of 346 aircraft dispatched. The RAF's losses on these operations were massive and unsustainable.

D-Day planners were right to be concerned about the Luftwaffe and the threat it posed to the invasion force. Its chain of day- and night-fighter airfields was supported by an extensive ground-based early warning radar network that stretched from Norway to the Pyrenees. Between Ostend and Cherbourg alone there was a major German radar station every 10 miles. However, the threat posed from the Luftwaffe by the time D-Day arrived was much less than imagined, both in respect of its ability to oppose the Allied air forces over Normandy and in terms of its own offensive capacity.

RIGHT *Freya and Würzburg-Riese* (Giant *Würzburg*) **radars near Le Havre, photographed from extremely low level by an RAF reconnaissance aircraft. Note the bomb damage to the building in the centre of the photograph.** *(Author's collection)*

The all-seeing eyes

The Germans were known to have an effective and well-integrated chain of radar stations along the invasion coast. With them they could gain early warning of the approach of Allied air and seaborne invasion forces, radar plots of the movement of Allied vessels, and accurate ranges and bearings for the control of their coastal guns. Although full integration of all constituent parts of the system was still several months from completion on D-Day, this formidable radar front presented an active threat to invasion operations.

Between Calais and Brest the actual count of German ground radars showed six *Wassermann* (Aquarian) and six *Mammut* (Mammoth) long-range early warning radars, 38 *Freyas* for medium-range early warning and night-fighter control, 42 *Würzburg-Riese* (Giant Würzburg) ground-controlled interception radar for night-fighter control, and coastal gun control for use against low-flying aircraft, and 17 *Seetakt* (Coastwatcher) and small *Würzburg* radars for gun laying, one per flak battery.

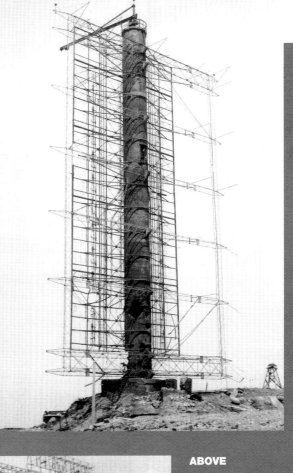

BELOW *Würzburg-Riese* was one of the most reliable and well-built radars of the war. Its parabolic dish was mounted on a purpose-built rugged four-wheel cross-section platform. *(USNA)*

ABOVE FuMG 404 *Jagdschloss* panoramic radar installation on a brick-built cabin. It comprised a rotatable stack of 16 horizontally polarised dipoles and reflectors mounted on a tower-shaped column seven metres high. Above this were mounted vertical broadband antennas for the transmission and reception of IFF signals on 125 and 156MHz. *Jagdschloss* was able to continue to operate with good results in the face of 'Window' (see caption on page 28 for explanation of 'Window') and other forms of jamming. *(Author's collection)*

ABOVE *Wassermann* antenna and rotating column at an installation in Denmark. There were a number of different versions of this long-range system along the French coast operating between 75 and 149MHz, and employing a vertical stack of *Freya* radar antenna arrays on a 4m-diameter 60m-high tubular steel mast. The *Wassermann L* and *S* versions had ranges of 200km and 300km respectively. *(USNA)*

RIGHT MTB 255 in floating dry dock. On her foredeck is a funnel-shaped high-frequency antenna, and on her mast-top is a low-frequency dipole antenna. She was destroyed in a mystery explosion in the harbour at Ostend on St Valentine's Day 1945 that killed her entire crew.
(R.G.R. Haggard)

Heavily armed night-fighter aircraft like the Messerschmitt Bf110, vectored on to their prey by ground-based radar, would cause carnage among the slow-flying formations of Dakotas and Stirlings crammed with paratroops, or towing troop-carrying gliders. Fighter-bombers like the Focke Wulf Fw190, let loose over the invasion fleet and the columns of troops struggling ashore on the landing beaches, could wreak havoc and cause heavy loss of life. By the time of D-Day the likelihood of these scenarios becoming a reality was small, but the Allies could still take no chances.

Therefore the planners needed to consider how best to prevent these doomsday scenarios from becoming realities. What they needed to do was conceal the passage of the invasion fleet for as long as possible. The answer was to create spoof amphibious and airborne landings to confuse the German defenders and dupe them into committing ground and air forces in areas of northern France well away from the actual landings.

The first step was to acquire detailed intelligence concerning the location and interconnection of enemy radar stations and networks, and technical details of the types, frequencies, characteristics and capabilities of individual radars.

A COSSAC (Chief of Staff to the Supreme

Allied Commander) committee, made up of representatives of all the armed services, began gathering the necessary information during 1943. When the SCAEF (Supreme Commander Allied Expeditionary Force) assumed the direction of the operation, this committee was carried on as a department of his staff. In the end, detailed and accurate intelligence became available, derived from photoreconnaissance, from radar monitors and from other sources. ANCXF (Allied Naval Commander Expeditionary Force) was given responsibility for operational policy and general orders, governing the use of Radio Countermeasures (RCM). In November 1943 a joint Navy–Air RCM plan was evolved that provided for the following measures to be taken to counter enemy radar:

- Accurate location of enemy coastal radars and their frequencies.
- Their destruction by air attack.
- On the eve of the invasion, electronic feint invasion forces would draw attention away from the real landing areas.
- Any radars still operational in the invasion area on D-Day would be electronically jammed.

The locations of the radar stations in northern France were triangulated using specially designed radio direction finders set up along the south coast of England, and then confirmed visually by photographic reconnaissance aircraft.

In a top secret operation coordinated by HMS *Flowerdown*, the Admiralty communications centre near Winchester, a specially modified Royal Navy MTB with a hand-picked crew were set the objectives of establishing the exact positions and frequencies of enemy radar sites along the Channel coastline, with particular emphasis placed on the gun-laying radars. By acquiring details of the radar frequencies, effective jamming could be put in place; and by knowing the locations of the radar sites the Allies could successfully deploy decoys to confuse the enemy about the likely location of the invasion that they knew was going to come.

MTB 255, belonging to the 14th MTB Flotilla based at HMS *Aggressive*, Newhaven, was fitted with a range of specialist radio equipment that included a high-frequency waveguide antenna and a low-frequency dipole antenna. The boat was stripped of any excess weight by the removal of her torpedoes and depth charges, and by only half-filling her fuel tanks, to reduce her draught sufficiently so she could pass over the top of any undetected inshore minefields. However, her gun was retained for defensive use.

The operations, codenamed 'Knitting', were carried out on very dark nights when there was little or no moon, and covered the enemy coastline from the Pas de Calais to the Channel Islands. Such was the importance attached to MTB 255's missions that she was given air cover, and two other MTBs would remain on station out at sea in case they were needed. MTB 255 carried out 23 'Knitting' operations that began in the months before D-Day and lasted until February 1945. The intelligence she gathered proved of great value to the Admiralty and the RAF.

Fighter-bomber Spitfires and Typhoons of the RAF's 2nd Tactical Air Force were given the difficult task of knocking out these well-defended sites. So as not to give the game away, for every 'real' target attacked they hit at least three similar targets elsewhere in France and Belgium. The anti-radar operations began on the morning of 16 March 1944 when rocket and cannon-armed Typhoons of 198 Squadron attacked the *Wassermann* early warning radar station near Ostend on the Belgian coast. The systematic destruction of enemy radar sites continued until the eve of D-Day, although a number still remained working on 6 June.

Operations Taxable and Glimmer

Allied planners sought to dupe the German High Command into thinking the invasion was going to take place either near Le Havre or Boulogne, convincing them to concentrate their forces in these areas.

Scientists at the British Telecommunications Research Establishment (TRE), Malvern, working with the American–British Laboratory Division 15 (ABL-15) countermeasures team, devised an ingenious piece of electronic trickery: they created two 'ghost armadas' that would move sedately across the English

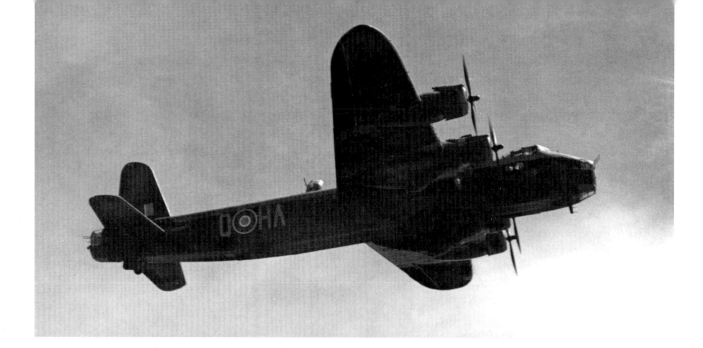

ABOVE No 218 Squadron's Stirlings were the Windowing force in Operation Glimmer. *(Author's collection)*

RIGHT A cloud of 'Window' floats down over occupied France. These strips of metallised paper (not unlike paper-chain strips in appearance before they are joined together) were cut to a particular length depending on the wavelength of the radar system to be jammed. Window was carried in bundles inside the aircraft. When untied it was pushed down the flare chute to be spread in the aircraft's slipstream and float to earth, fogging the enemy's radar screens with spurious returns. *(Author's collection)*

Channel towards different points on the French coast at a speed of seven knots – and it was aircraft and not ships that were to create this spoof invasion force.

The larger of the two fleets, codenamed Operation Taxable, covered an area 14 miles wide and 16 miles deep off the coast of Upper Normandy between Fécamp and Le Havre, and was set to last 3½ hours in the early hours of 6 June. Eight Avro Lancaster bombers from the famous 617 Squadron (the Dam Busters) were to fly carefully planned tracks over the sea while dropping large quantities of 'Rope' – long lengths of the radar-reflective foil codenamed 'Window' – to jam the coast-watching *Seetakt* radar that operated in the 370MHz band. Precision flying was needed in order to drop the required pattern of Rope clouds that would produce a continuous blip on German radar screens, thereby simulating an invasion fleet moving slowly forward. As a contingency against equipment failure, each Lancaster carried two Gee sets (a radio navigation system) as well as two navigators, and four men to drop the Window.

Operation Glimmer was a smaller spoof further up the French coast. It involved six Short Stirling bombers of 218 Squadron flying an identical track pattern to Taxable while also dropping Rope, and was aimed at the coast between Dunkirk and Boulogne. Each aircraft was equipped with a Gee and a Gee-H navigational system with three navigators to monitor the tricky flying patterns of the operation, and four men to dispense the Window.

To lend credence to the deception, four RAF air-sea rescue (ASR) launches were assigned to each operation, each of which carried a Moonshine electronic signal repeater. Moonshine picked up signals from the *Hohentwiel* radar carried in Luftwaffe maritime reconnaissance aircraft then amplified and re-transmitted them, giving the impression of a massive force of ships sailing close together.

Each ASR launch, in company with 14 small naval launches, towed a raft-like float flying a 29ft-long naval barrage balloon (Filbert) containing a 9in radar reflector inside its envelope, which gave a radar echo like a large ship. These vessels cruised on the waters of

Racecourse pattern

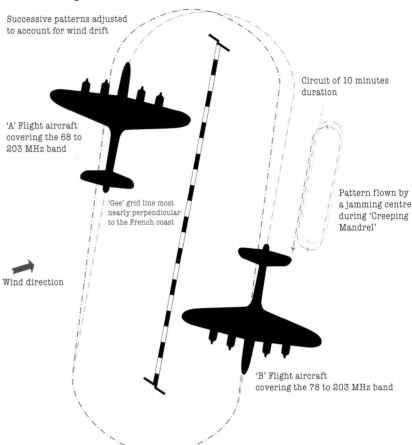

Successive patterns adjusted to account for wind drift

Circuit of 10 minutes duration

'A' Flight aircraft covering the 68 to 203 MHz band

Pattern flown by a jamming centre during 'Creeping Mandrel'

'Gee' grid line most nearly perpendicular to the French coast

Wind direction

'B' Flight aircraft covering the 78 to 203 MHz band

the English Channel beneath the falling Rope to make the hoax even more convincing.

Sixteen Stirlings from the RAF's 199 Squadron, and four B-17 Flying Fortresses of the 803rd Bombardment Squadron on loan to the RAF from the USAAF, set up a Mandrel screen in a line from Littlehampton to Portland Bill to give electronic cover to the advancing invasion fleet, but with the jamming deliberately thin to the east to allow the German operators to observe the Taxable and Glimmer spoofs.

Beneath the orbiting aircraft and their falling clouds of Window, the small flotilla of naval launches, with their ungainly Filbert balloons trailing downwind, headed south into the unsettled waters of the Channel.

Meanwhile, five B-17s of 214 Squadron joined forces with 22 Lancasters of 101 Squadron to fly a 'ghost' bomber stream heading into France, dropping large quantities of Window and using their 'Airborne Cigar' jammers along the line of the Somme estuary.

ABOVE The 'racecourse pattern' used by the RAF in Operations Glimmer and Taxable.

(Dominic Stickland)

Its purpose was to distract enemy night-fighters away from the formations of RAF and USAAF transport aircraft heading to their drop zones at Sainte-Mère-Église and beside the River Orne.

No 214 Squadron's Fortresses were airborne for 6½ hours, patrolling back and forth at 27,000ft along a 90-mile line, and 101 Squadron's Lancasters were airborne for up to 7½ hours, patrolling from 11:51pm to 4:34am between 20,000ft and 26,000ft. Of the 22 Lancasters dispatched, six abandoned their missions because of technical problems with aircraft and equipment, and in one instance because of airsickness of a crewmember.

The airborne jamming was augmented by ground stations in Britain, and the Mandrel screen, Ground Mandrel and Window were all targeted at enemy early-warning radars in the 90 to 200MHz band. The Window used was Type MB, designed to jam the German *Freya* radar system.

The Mandrel screen employed what was termed a 'racecourse pattern', where aircraft flew up and down the Gee lines that were most nearly perpendicular to the enemy coast. An unwelcome own goal of the Mandrel jamming was that it adversely affected Gee transmissions, which meant that navigators needed to use dead reckoning to calculate wind drift once the screen was in operation. The first circuit was 10 miles long, with rate-one turns being made between the straight legs, and every circuit that followed was modified to compensate for the effects of the prevailing wind conditions so that it took precisely 10 minutes.

Each pair of aircraft formed a 'jamming centre', with each individual aircraft starting at the opposite end of the circuit and chasing the other aircraft, thereby 'maintaining a strong average intensity of jamming'.

Smokescreens

A far simpler and more cost-effective means of concealing the invasion fleet was the use of smokescreens. On D-Day the great importance given by the British to concealment using smoke was evidenced by the involvement in the task of both the Royal Navy and the RAF. US commanders had made plans for the use of large area smokescreens by chemical battalion troops using portable smoke generators during the assault phases on Omaha and Utah beaches, but in the event they decided it was not necessary to use smoke because the threat from enemy air attack was low.

Prior to D-Day the USAAF had carried out trials with P-51 Mustang and P-47 Thunderbolt fighter-bombers fitted with under-wing M10 smoke tanks to lay smoke curtains 2,000ft long, but in the end this system was not used in Normandy either.

On 6 June Royal Navy Landing Craft Personnel (LCP) provided smoke cover for the assaults on Sword, Juno and Gold Beaches, and for some of the bombarding warships lying offshore. To 'make smoke', chlorosulfonic acid (a highly dangerous substance) was forced under pressure through a jet at the LCP's stern and reacted on contact with the air to produce dense clouds of white smoke.

Taxable navigator
Flg Off Thomas Bennett, Lancaster bomber navigator, 617 Squadron RAF

When the operation order came through it had been decided that eight aircraft would go off in the initial phase and would be relieved after two hours by the second eight... We had to give the second wave start times on the English coast and they would fly down and pick up the aircraft they were relieving and do their last circuit with them. The aircraft that was Windowing would be at 3,000ft; the relieving aircraft would come along at 3,500, pick them up and do their last circuit with them.

The first aircraft would then fly off back to base and in the turn the second aircraft would drop to 3,000ft and pick up the sequence. But this had to be done within a tolerance of 90 seconds, otherwise the whole of that segment of the convoy would disappear from the radar screens, and of course that would mean the Germans would immediately suspect the veracity of this thing. So this was something that concerned us very greatly but in the event the handover, the takeover, went off beautifully.

The RAF supplied two squadrons of Douglas Boston Mk III medium bombers fitted with Smoke Curtain Installation (SCI) canisters inside their bomb bays. Each aircraft carried four SCI canisters, with each canister capable of projecting 11 seconds' worth of white smoke into the atmosphere via four dispenser pipes protruding downwards through the bomb-bay doors.

Operating from RAF Hartford Bridge in Hampshire, 88 Squadron's Bostons covered the eastern (British and Canadian) flank of the invasion fleet, while those of 342 (Free French) Squadron took the west (American beaches). It was a precision operation with successive relays of Bostons, two at a time, making their approach at 300ft to arrive on each flank at 10-minute intervals (the first at 5:00am) in order to maintain a continuous smokescreen. Each aircraft dived to sea level to lay its smoke curtain that could be up to 5,000yd long. There is no doubt that the effectiveness of this screen played an important part in the success of the Normandy landings.

Operation Titanic

Soon after midnight on 6 June a mixed force of 40 RAF aircraft – comprising Lockheed Hudsons and Handley Page Halifaxes from 138 and 161 Squadrons, and Short Stirlings from 90 and 149 Squadrons – mounted fake airborne assaults over Normandy. Codenamed 'Titanic', the execution of the mission was in four parts and called for the dropping of 500 dummy parachutists, rifle fire simulators, Window and two SAS teams to simulate airborne landings away from the actual invasion area and paratroop drop zones.

Titanic I   The simulated drop of an airborne division north of the River Seine near Yvetot, Yerville and Doudeville in the Seine-Maritime Départment, and at Fauville in the Eure Départment.

Titanic II   Dropping 50 dummy parachutists

ABOVE A smoke-laying Boston of 88 Squadron flies low over the English Channel towards the invasion area. Note the smoke dispenser pipes beneath the fuselage. *(IWM FLM 2585)*

BELOW Short Stirlings of 149 Squadron flew fake 'invasions' in Operation Titanic. *(Author's collection)*

east of the River Dives to draw German reserves on to that side of the river (the mission was cancelled shortly before 6 June).

Titanic III Dropping 50 dummy parachutists in the Calvados Départment in Lower Normandy near Maltot, and in the woods to the north of Baron-sur-Odon, to draw German reserves away to the west of Caen.

Titanic IV Dropping 200 dummy parachutists near Marigny, west of Saint-Lô in the Manche Départment, to simulate the dropping of an airborne division (as with Titanic I) to the south of Carentan. Two sticks of six men from 1st Special Air Service Regiment (SAS) were also dropped near here at 12:20am on 6 June. The SAS teams carried sound recordings and amplifiers and replayed the sounds of rifle and mortar fire along with shouted commands. The recordings lasted for 30 minutes, after which the teams withdrew from the area.

The 12-man SAS team was commanded by Captain Harry Fowles and Lieutenant Noel Poole. They had been given orders to allow some of the enemy to escape to spread alarm by reporting landings by hundreds of parachutists, but they lost their real weapons containers in the jump and were forced to hide away from the Germans. They did so for a month before they were tracked down and captured. Eight failed to return from the operation, either killed in action or executed by the Germans later.

From the seven aircraft dispatched by 149 Squadron on Titanic III, two Stirlings and their eight-man crews were lost to unknown causes in this operation.

As dawn approached along the Normandy coastline on 6 June, operators of the few surviving coastal radars reported the approach of the Glimmer ghost fleet, and German High Command issued an alert to expect landings in the area of Calais and Dunkirk. More than five hours after Glimmer ended the Germans were still searching the waters off Dunkirk and Boulogne looking for the reported invasion force. Operation Taxable drew less attention than Glimmer, probably because the fighter-bomber attacks by 2nd Tactical Air Force had been too effective. There were very few functioning radar installations left in the area, so the approaching ghost fleet could not be seen.

Operation Titanic had also been a success. The Germans were tricked into believing parachutists had landed east of Caen and further to the west in the countryside around Coutances, Valognes and Saint-Lô at about 2:00am and the German 7th Army was placed on full invasion alert. Part of the 12th *SS Panzerdivision Hitlerjugend* were ordered at 4:30am to deal with a parachute landing on the coast between Houlgate and Trouville, which was found to be dummies from Titanic III. The unintended scattering of the American 82nd and 101st Airborne Divisions in the Cotentin, as well as elements of the British 6th Airborne around the River Orne, had an unexpected (but positive) outcome for the Allies. The German 352nd Infantry Division's *Kampfgruppe Meyer* was sent on a pointless wild goose chase to round up these paratroopers when they could have been used more effectively to intervene on the landing beaches at Gold and Omaha. Enigma intercepts from the area of Titanic I revealed that the local commander had reported to the German High Command a major landing up the coast from Le Havre.

BELOW Dummy parachutists were a part of the 'invasion' in Operation Titanic. *(Author's collection)*

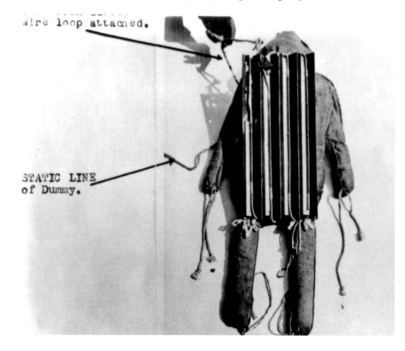

wire loop attached.

STATIC LINE of Dummy.

Maritime electronic navigation systems and radar

On 5 June 1944, 21 minesweepers and other vessels were fitted with a secret device codenamed 'Admiralty Outfit QM', which enabled them to accurately navigate across the English Channel and sweep the minefields in the planned landing areas, opening up safe channels through which the invasion armada could pass to the landing beaches.

One of the major Allied naval tasks of Operation Neptune was to protect the armada of ships and seaborne invasion forces from the danger of enemy minefields. The Allied armada was to sail from numerous ports along the south-west, south and south-east coasts of England and was then to converge on Area Z (a circular assembly area in the Channel, 10 miles in diameter), before turning southward and crossing the Channel through 'the Spout' to approach the assault area along one of five swept sea lanes, which corresponded with the landing beaches.

The requirement of the naval minesweeping plan was to ensure that those routes, the assault area anchorages and the manoeuvring space for vessels were free of enemy mines. However, the difficulty of navigating a minesweeper across the English Channel and making a precise landfall at night was considered impossible without some form of accurate radio navigation. The solution lay in the use of two similar radio navigational systems – Gee (Outfit QH) and Decca (Outfit QM). These enabled vessels to fix their positions to a high degree of accuracy by day or by night, and in all weathers.

It is important to note that the first of these hyperbolic navigation systems (Gee) was in widespread use at the time with the RAF and Royal Navy. The naval version of Gee was first used in the ill-fated Dieppe raid in August 1942, and was subsequently established as a standard system for surface navigation. For Operation Neptune the initial legs of swept channels were planned to coincide with the same lines as the Gee lattice maps. So important was accuracy in this task that some 860 invasion ships were fitted with Gee.

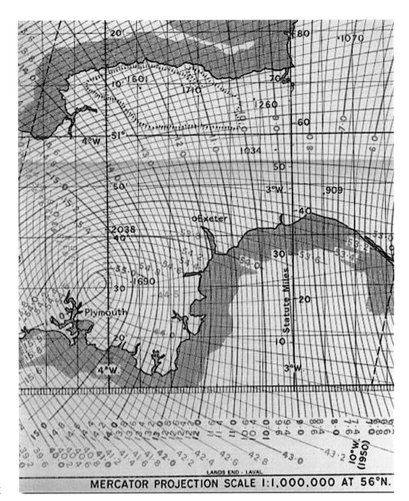

MERCATOR PROJECTION SCALE 1:1,000,000 AT 56°N.

Gee (Outfit QH)

Gee was the codename given to the radio navigation system used by the RAF, devised in the late 1930s by Robert J. Dippy at Robert Watson-Watt's radar laboratory at RAF Bawdsey, and further developed at the Telecommunications Research Establishment at Swanage. It consisted of three transmitters (one master and two slaves) located at known positions in southern England, each transmitting a 'pip' simultaneously. The differences in time in which the pips from the two transmitters (the master and either of the two slaves) were received established a line of points along which the receiving ship could be located. The difference in time at which another pair of pips was received established another line along which the receiving ship could be located. The point at which these two lines intersected was the location of the ship. Gee took its name from the two-dimensional grid generated by the two pairs of stations, Gee being the 'G' in 'Grid'.

ABOVE Gee lattice grid. *(Copyright unknown)*

Decca (Outfit QM)

One of the Admiralty's wartime trials of the Decca system almost ended in disaster. For ease of use, the ship performing the evaluation was navigated on a single hyperbolic line, which passed through a reef. Intent only on watching the decometers, the ship's navigator steered exactly on the line and nearly ran the vessel aground, not thinking that Decca could be so incredibly accurate.

For Operation Neptune, four Decca (Outfit QM) transmitting stations were established in great secrecy along the south coast of England. Intense security surrounded their construction because knowledge of their location could reveal to the enemy the sites of the invasion beaches. The master station, known as 'A', was built near Chichester, the western 'B' (Red) station near Swanage and 'C' (the Green slave) about a mile inland from Beachy Head. In the Thames estuary on the Isle of Sheppey a transmitter was built to look like a 'decoy' in case the Germans discovered any part of the plan. (After the war it became usual to establish a fourth (Purple) transmitter for additional accuracy, but it was not considered necessary for the purpose of D-Day.)

Signals from the Red and Green transmitters formed hyperbolic patterns, known as lines of lattice, which were plotted on maps. ATS plotters who calculated the Decca lattice lines worked in pairs in a hut at the ASE, under armed guard, so secret was their work.

The Decca Company made 20 pre-production Decca receivers at very short notice and were fitted into the lead vessels of the 13 British, American and Canadian minesweeping flotillas, five headquarters landing craft (LCH), and two navigational motor launches.

Transmissions from the chain began early on 5 June. William J. O'Brien, the American engineer who developed Decca Navigator, kept a prototype receiver turned on in his London home. When the decometer dials came to life he knew that the invasion was under way. There were also monitoring receivers set up on shore.

Type 970 Radar

Certain Motor Launches (ML) were designated navigational leaders for the minesweeping flotillas and the LCT flotillas on the run-in shoot (see page 54). They were equipped with Type 970 Radar. A projector was fitted to the plan position indicator (PPI) of this radar, from which a slide (similar to a photographic slide or transparency) could throw an image on to the PPI scope. Predictions were prepared which cast an image on the PPI scope exactly like that which the radar would show when the ship was located at a certain spot. When the radar image and the predicted image were in register and matched, the ship was known to be on the spot indicated. Predictions were made for points on the line of approach 5,000yd and 1,500yd (the line of departure) off-shore. Using these predictions, the navigational leaders brought the landing waves and control vessels to the intended line of departure with a high degree of accuracy.

When it works, technology can have tremendous benefits, but when it goes wrong (for whatever reason) the consequences can be unexpected – and in a war situation, potentially disastrous. On D-Day there were some critical failures of radar in assault craft as well as a great many successes. At Pointe du Hoc on Omaha Beach and in the Utah sector, failures of navigational equipment had some grave implications for the American landings.

BELOW Decca lattice grid. *(Dominic Stickland)*

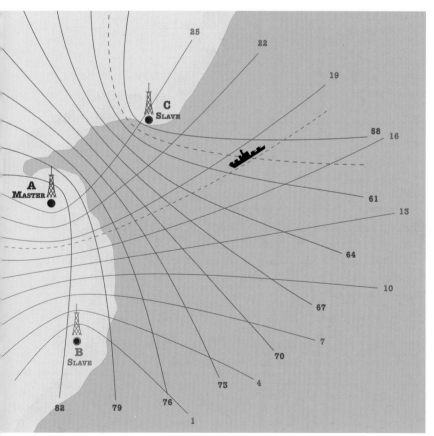

Gee is for genius

English civilian research scientist Robert 'Bob' Dippy is one of the great unsung heroes of D-Day. He was the brains behind the radio navigation aids known as Gee and Loran, which made it possible for hundreds of ships to find their way safely along suitable Gee grid lines through mineswept channels off the Normandy coast; and for massed formations of aircraft to fly confidently in cloud and darkness to their drop zones over France, blind except for Gee. Without these devices, D-Day could easily have descended into chaos.

The son of a naval architect, Robert James Dippy was born at Charlton, Kent, on 26 November 1912. He was educated at Dartford Grammar School and at the University of London from where he graduated in 1934 with an honours degree in electrical engineering.

Dippy joined the Research Laboratories of the (British) General Electric Co. in late 1934 and was soon engaged in television development. He responded to a newspaper advertisement for positions in radio research in the Air Ministry and found himself appointed in July 1936 as one of the 'Foundation Members' of the Bawdsey Research Station in Suffolk – the birthplace of British radar, later to become the Telecommunications Research Establishment. There, he worked on radar development as part of Robert Watson-Watt's team.

In about 1938 Dippy suggested a radar-like system to aid approach and landing at airports in bad visibility. Owing to Bawdsey's urgent work on Chain Home radar and airborne systems, his proposals were not immediately acted upon, but Dippy brought it forward again in 1941 at a time of critical need by the RAF. The device (a pulsed hyperbolic system for radio navigation), that became the Gee system of navigational guidance, made the RAF's first 1,000-bomber raid of 1942 practicable, and made Gee equipment the most widely installed of all radar and radar-like airborne equipment – vastly improving the RAF's navigational and bombing accuracy. Many bomber crews had Gee to thank for getting them safely home to their bases in all weathers.

In the summer of 1942 the Royal Navy had begun to install Gee in their light surface craft, enabling them to navigate accurately in any weather. Minesweeping and mine-laying were also Gee-aided operations. In fact, Allied air and sea forces used Gee so extensively on 6 June 1944 that it has been said that D-Day was G-Day!

ABOVE Bob Dippy was the brains behind the invention of Gee. *(Author's collection)*

Gee and Loran went through further development as Gee-H, and in the USA (with Dippy as adviser) as Loran-A. Eventually the use of Gee was widespread, and Dippy was subsequently awarded the OBE (1948) and the American Medal of Freedom with Bronze palm. He was also given a departmental award of £5,000 in recognition of his work.

After the war, Dippy became Senior Principal Scientific Officer at the Ministry of Supply before migrating to New Zealand in 1949 to join the Civil Aviation Branch of the Air Department as Controller of Telecommunications. By 1957 Dippy was feeling the urge to return to scientific research and went to Australia to join the Weapons Research Establishment near Adelaide as Principal Scientific Officer, Electronic Techniques, progressing to Superintending Scientist in charge of the Applied Physics Division. He received the Pioneer Award of the IEEE in 1966 for hyperbolic radio navigation.

On retirement, Dippy returned to England and died on 3 July 1984 at Folkestone, Kent, aged 71.

Operation Enthrone and sonic buoys

To ensure that the channels to be swept could be relocated with ease on D-Day, 10 sonic underwater buoys (FH-830 Acoustic Beacon) were laid on D-6 at the northern end of the 10 approach channels (two for each assault force) to the landing beaches. They were laid sonically 'dead' but with a timer to 'come alive' on D-1, when they would be used by Harbour Defence Motor Launches (HDMLs) acting as marker boats to show the minesweepers where they should begin their sweeping of the approach channels.

All key marker buoy HDMLs were equipped with special radar reflectors, which gave a distinctive response on ships' radars. SG (seagoing) surface-to-surface radar was installed in about one in six Infantry Landing Craft (LCI) and larger landing craft, so the reflector buoys were a great safety asset. Thus ships were able to take bearings whether they could see the buoys or not.

Ten flotillas of nine fleet minesweepers each were to cut the ten channels and to search the five 'transport areas'. Attached to each flotilla were four minesweeping motor launches, two Oropesa minesweeping 'LL' trawlers, and four danlayers. Equipped with light sweeping gear, the motor launches preceded the leading fleet sweeps to clear a path for them. The LL trawlers swept for magnetic mines while the danlayers marked the swept channel with buoys to guide the oncoming invasion fleet. The total strength of Allied vessels engaged in the minesweeping operation was 255.

RIGHT Decca mine-swept channels for Operation Neptune. *(Dominic Stickland)*

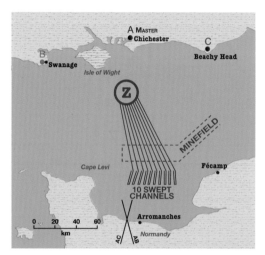

RIGHT Crew work the drum on board a Royal Navy minesweeper to stream the LL sweeping gear used to detonate magnetic mines. The LL sweep consisted of two unequal lengths of buoyant cable with electrodes towards their tail ends, trailing behind a minesweeper. These cables were powered by a series of pulsating electrical charges, which in turn induced a magnetic field strong enough to fire the mine. *(IWM A15551)*

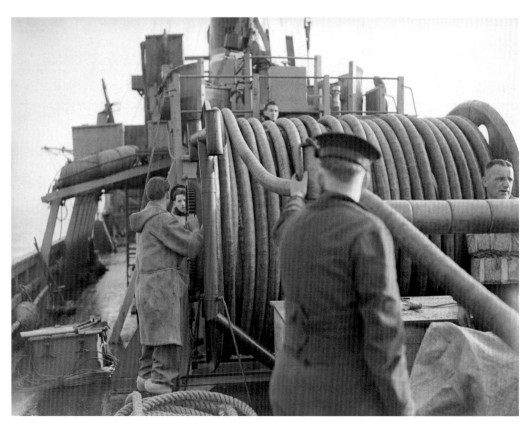

Accuracy

Once sweeping was begun from the correct starting points, accuracy of navigation was aided by the use of QH and QM and by taut wire. Thanks largely to the use of these two radio devices, the Royal Navy's 18th Minesweeping Flotilla, for example, sailing from Portsmouth, was only about four minutes late and 400yd out of position on reaching the destination point off its designated beach, despite encountering strong winds and tides during the passage. This was considered a reasonable margin of error. One of the navigation officers is reported to have said afterwards: 'It was uncanny. It seemed as if we had some sort of overhead cable, which not only showed us the direction but also our speed.'

Once the sweeping of the approach channels and transport areas was completed, the minesweeping flotillas turned their attention to clearing lateral channels connecting the inshore extremities of the ten approach channels, and then progressively widening the approach channels to make a single broad channel into each Task Force sector.

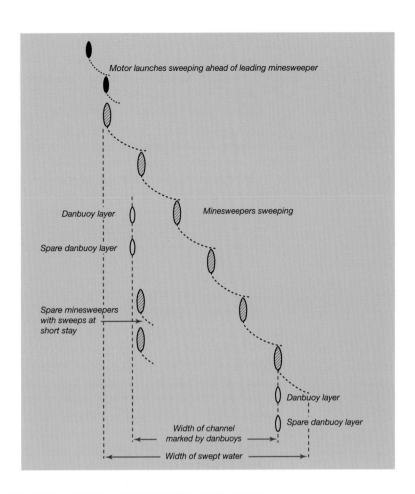

ABOVE **Minesweeping schematic.** *(Dominic Stickland)*

Allied ships lost to enemy mines, and mines accounted for, 4 June–3 July

	Western Task Force	Eastern Task Force	Total
Casualties	24	19	43
Moored mines swept	91	95	186
Ground mines swept	140	109	249
Ground mines accounted for other than by minesweepers	6	68	74
Total mines accounted for	237	272	509

Note: The Germans could have massively increased the mine threat to Allied vessels on D-Day had they started laying their new generation of pressure-activated sea mines in the weeks before 6 June. Had they done so, there was very little Allies could have done to detect or counter them.

The clearance of the waters between Channels 3 and 4 (known collectively as Channel 34) and between 5 and 6 (Channel 56) went according to plan and was completed on D-Day. Channels 12 and 78 were completed on D+1; Channel 14 was ready on D+7; and Channel 58 was cleared to a width of six miles by D+8. The entire barrier within the SPOUT was open to sea traffic by D+12. Clearance of the enemy mine barrier was extended far enough north to ensure that the whole of the minefields discovered during the approach was cleared. Some 78 moored mines were found in this field alone.

Gee or Decca?

In Operation Neptune, Gee was found to be accurate to within 100yd and Decca to within 50yd. But, invaluable as they both were, Gee and Decca also had their shortcomings. Although both systems were similar in broad principles, Decca was more accurate and more user-friendly because its readings were presented directly on instrument dials called 'decometers' instead of a cathode ray tube as used on the Gee. But one disadvantage of the early Decca sets was the need to set up the decometers using an accurately known position. If there was a break in reception for any reason the decometers had to be recalibrated. Although the Decca system worked extremely well and provided greater accuracy than Gee, it was less reliable because of a problem known as 'lane slipping'. However, ships fitted with Gee receivers could only navigate accurately as far as the mid-Channel mustering point.

To the chagrin of those on the mine-sweepers, who could have made good use of it, the Decca chain was switched off on D+1, presumably because the system was so secret at the time. Some thought it was shut down because of transmitter troubles. Decca was never jammed and its existence during the war was probably never suspected by the Germans. Perhaps because it was so secret, but maybe because so few Decca sets were fitted compared with Gee, the use of the Decca navigator system in Operation Neptune is rarely mentioned in any published accounts of the invasion.

Fighter Direction Tenders

A large number of Allied aircraft was expected to be operating in the airspace over the Normandy beachhead, so a command and control system to manage them was considered vital. To this end Landing Ship, Tank (LST) vessels were converted into Fighter Direction Tenders (FDT) carrying specialised radar equipment. The FDT, under naval command with an RAF directing crew, were sailed into position on D-Day and operated off the Normandy coastline during the assault phase to control Allied aircraft and intercept hostile aircraft until shore-based radar could be set up in France.

The concept of operating ground radar equipment on specially converted sea-going vessels was experimented with during the Allied invasion of Sicily in July 1943 when an RAF mobile Ground Controlled Interception (GCI) radar unit was mounted on an LST. The results were so successful that the C-in-C Fighter Command recommended that similar vessels should be used in the landings in northern France planned for 1944.

Three LST were chosen for conversion to FDT configuration by John Brown & Co at Clydebank, and the work was finished in February 1944. The remodelling involved welding shut the bow unloading doors and covering the hatches with armour plate. Because of the LST's shallow draught, some 300 tons of pig iron were secured to the main deck to slow the rolling action of the vessel, and a new deck was laid inside over the vehicle cargo space to contain personnel accommodation for the Filter Room, Communications Office, Cipher Office, Air Control Room, and Radar Receiving Room. A Direction Finding Office was constructed in the forward area for the installation of naval D/F equipment, while space was provided at the aft end of the tank deck for the Transmitter and Transceiver Rooms, Aircraft D/F, Radio Countermeasures Office and a W/T Storeroom. A Bridge Visual Direction position and a Bridge Plot House were constructed above deck.

FDT successes 6–13 June 1944

	A/c controlled	Contacts obtained	Friendly a/c intercepted	Enemy a/c destroyed/damaged
FDT 216	62	49	33	3/0
FDT 217	225	66	37	3/2

The all-important radar equipment in each FDT comprised a Type 15 GCI (forward) and a Type 11 GCI (amidships), the aerial arrays for which towered 30ft above water level. Mk III IFF (Identification Friend or Foe) was installed to provide identification only when the Air Movements Liaison Section information did not provide the answer. Airborne Interception (AI) beacons were installed to assist in the control of night-fighters, each FDT carrying 1.5m Mk IV and 10cm Mk VIII AI beacons. To provide voice communication between FDT and aircraft, multi-channel VHF R/T and W/T were fitted.

Once converted to the FDT role, each vessel had a displacement of 3,700 tons and a length of 328ft, roughly comparable in size to a Royal Navy destroyer. It carried a Navy crew of eight officers and 92 ratings, with 19 RAF and RCAF officers and 157 airmen.

The three FDT controlling RAF and USAAF fighters during the assault were:

■ FDT 216 – in the western half of the assault area off the American beaches Omaha and Utah;

■ FDT 217 – in the eastern half of the assault area off the British and Canadian beaches;
■ FDT 13 – in the main shipping route.

At 10:00pm on 5 June, the three FDT set sail from Cowes on the Isle of Wight to join the Assault Task Force. All three vessels commenced radar watch at H-Hour, 7:25am on 6 June, and were immediately in contact with the Allied fighter aircraft providing air defence to the Task Force.

On 15 June, FDT 216 returned to port for repairs and was relieved by FDT 217 off the American beaches; FDT 13 was positioned 20 miles east-north-east of Barfleur to cover the approaches to Cherbourg. FDT 217 was withdrawn after 17 days of continuous operation on 23 June. On 27 June, FDT 216 relieved FDT 13 off Barfleur, and on 1 July, when the shore-based GCIs were working inland from Barfleur, FDT 216 was moved up the coast 23 miles west of Le Havre. On 7 July she was hit by an aerial torpedo fired from a Junkers Ju88 and capsized. All but five of her crew were rescued. FDT 13 and 217 survived the war.

BELOW Fighter Direction Tenders (FDT) were floating command and control centres bristling with radar and radio antennae, with sophisticated communications equipment below decks, which were positioned off the Normandy beaches. FDT 216 was one of three such vessels that had been converted from Landing Ship, Tank. *(IWM A21922)*

Chapter Two

On enemy shores

Landing ships and craft

The daylight landings on the Normandy coast called for a wide variety of landing craft types. Many new weapons were also used operationally for the first time, which made the planning and implementation of the assault a complex affair. Some 3,286 landing craft of all sizes were assigned to Operation Neptune, comprising 25 different kinds of vessel.

OPPOSITE Royal Navy Landing Craft, Assault (LCA) 1377 on manoeuvres in Weymouth Bay before D-Day. At 7:30am on 6 June, LCA 1377 and its Royal Navy crew landed men of the US 5th Ranger Infantry Battalion on Omaha's Dog Green sector, the first landing craft to land on that section of Omaha Beach. The 'PB1' markings on the LCA's flank designate No 1 landing craft assigned to HMS *Prince Baudouin*. *(USNA)*

RIGHT The Landing Ship, Tank (LST) was among the first of the new breed of assault vessels. It was the Royal Navy who inaugurated LST construction in the USA in 1942 when they placed an order for 200. Seen here, just before D-Day, LST 357 is loaded at Portland harbour in Dorset with US Army vehicles destined for Omaha beach. In the foreground is a Dodge WC-51 4 x 4 weapons carrier, and on the right a GMC AFKWX 353 (DUKW). The stork motif inside the bow doors of the LST bears the legend 'we delivered'. By D-Day 357 was an old hand. Nicknamed 'Palermo Pete' by her crew, she had already taken part in the invasion of Sicily in July 1943, and later that year in the landings at Salerno. She survived the war to be scrapped in 1948. (USNA)

ABOVE Amphibious assault, Gallipoli, 5 May 1915: soldiers of the Lancashire Fusiliers on board an old Royal Navy battleship used in the third phase of operations in the Dardanelles Straits before they disembarked at 'W' and 'V' beaches off Cape Helles. *(Author's collection)*

The landings on the beaches of Normandy on D-Day were the largest amphibious assault in history. Almost 30 years earlier, during the First World War, British and ANZAC troops were landed at Gallipoli from open rowing boats on to a ferociously defended shore, suffering horrendous losses as they attempted to push inland. Thankfully, the D-Day planners heeded the lessons of that catastrophic failure, and by the outbreak of the Second World War military thinking and assault craft design had come a long way. When Operation Overlord was unleashed on 6 June 1944, Allied troops were landed on the Normandy beaches from an array of purpose-built landing craft that were very

much up to the tasks required of them.

After the fall of France in June 1940, Prime Minister Winston Churchill was intent on projecting British power back on to the continent of Europe. The tools for this job were landing craft and ships to support commando operations, but Britain's shipbuilders were already being stretched to breaking point, so Churchill looked to the US to supply the vessels he so urgently needed. British-inspired designs like the LST (Landing Ship, Tank), LSD (Landing Ship, Dock), LCI(L) (Landing Craft, Infantry (Large) and the LCT (Landing Craft, Tank) were responsible for some of the most important wartime American classes. Britain was also

RIGHT Amphibious assault, Normandy, 6 June 1944: this photograph was taken from a tank landing craft (LCT) as it approached the beach in the Queen Red sector of Sword beach at about 8:00am on 6 June. The smoke and fury of the assault are plain to see on what was among the bloodiest sectors of the British landing beaches. *(IWM B5111)*

The Landing Craft, Infantry (LCI) comprised several classes of ocean-going amphibious assault ships, developed by the US in response to a British request for an assault vessel capable of carrying more troops than the Royal Navy's smaller LCA. Displacing 389 long tons laden, the LCI(L) was 158ft 6in long and 23ft 3in in the beam. It could land its cargo of 210 troops from the deck using two ramps located either side of the bow. LSI(L) 84 (seen here) was also a battle-seasoned assault vessel. She had seen service in Tunisia, Sicily and Salerno in 1943 before taking part in the Normandy landings. Some 250 LCI were used by the US Navy and the Royal Navy in the Normandy landings. *(USNA)*

the main instigator behind the concepts of the amphibious headquarters ship, amphibious control craft, and close-in fire support craft. American practice was heavily influenced by British experience and doctrine, particularly in the Atlantic theatre.

For most of the Second World War, the shortage of landing craft was a recurrent theme in Allied strategic planning. The situation became critical in the summer of 1942 when the US Army was told to prepare for the invasion of France. Because of the distance across the English Channel between Britain and France, the US Army wanted enough landing craft to lift its whole expeditionary force in one single trip without having to return to England for a second journey. But the big question was whether American and British boat and shipbuilders could deliver on the big requirement for assault vessels.

The main problem facing the provision of landing craft and ships was that there was no long-term construction strategy in place. Instead, vessels were being turned out as and when they were required for particular operations. In the case of small landing craft,

this lack of a system was not so much an issue, but the larger landing craft took more time to build, and the even bigger LST needed a six-month build time initially.

Much emphasis was placed on the provision of large numbers of small craft, upon which the delivery of assault troops to a beachhead depended. Requirements for the amphibious invasion of France across the English Channel continued to grow.

Speaking after the war about the procurement of landing ships and landing craft, Vice-Admiral Lord Louis Mountbatten, Chief of Combined Operations (1941–43) had this to say:

The absolutely crucial thing for an invasion is to get the troops across the water, and for that you want landing ships and landing craft, and those we just didn't have. They had to be designed; they had to be built in large quantities at a time when all the shipbuilding facilities were required to fight the Battle of the Atlantic. But we managed to get permission to get smaller yards to start building the landing craft, and then we started converting merchant ships to

landing ships. And above all, when the Americans came in, I persuaded General Marshall right away to double all the orders I'd placed in America. That's how we built up the landing craft at a time when nobody wanted them to be built.

(Britain built only 24 landing ships in the entire six years of war, 21 of them in 1945, which by then was too late to matter. But what it did build was 1,264 major landing craft and 2,867 minor ones; US production totals were 1,573 landing ships, 2,486 major craft, and 45,524 minor craft.)

In July 1942, plans for the French operation were put on hold, with attention being redirected towards North Africa and planning for Operation Torch, the Anglo-American invasion of French North Africa. Torch was delayed for several months (until November) owing to a shortage of assault transports and landing craft, and when the landings finally took place these shortages meant that not all the objectives were achieved. At the same time American losses in the Pacific needed to be made good with new ships and landing craft.

In 1943 and 1944, Allied military operations around the world continued to be hamstrung by a lack of assault vessels. Orders for landing craft in the US took up the whole capacity of the nation's boat-building industry, but the weak link in the chain was the supply of engines to power the craft. On occasions the tough decision had to be taken on whether diesel engines should be allocated to armoured fighting vehicles or to landing craft.

At the time of Overlord in June 1944, most of the landing craft in Allied service were built in the USA. The exception to this was the British-designed and built Landing Craft Assault (LCA), which was the most numerous British and Commonwealth landing craft of the Second World War. In the months leading up to the invasion, British shipyards continued to be heavily involved with servicing the needs of the Royal and Merchant navies in the Battle of the Atlantic – repairing war-damaged vessels, building convoy escorts and merchant ships; and from late 1943 onwards making components for the Mulberry harbours. So, there was little in the way of spare capacity for building a range of different landing craft.

Spoilt for choice

A bewildering variety of landing ships and craft were used on D-Day – more than can be fully covered in this book, as it is a subject that deserves a book of its own. Largest of all was the LST, of which 236 were used in the assault, carrying tanks and specialised armour, together with soft-skinned transport and armour. They were supported by 837 LCT, the smaller of the tank-landing vessels. Infantrymen were carried in a number of different craft, ranging from attack transports to LCP

Landing ships and craft

Below is a list of commonly used abbreviations for just some of the vessels operated by the US Navy, the Royal Navy and Commonwealth navies on and after D-Day:

DUKW	'Duck' amphibious truck
FDT	Fighter Direction Tender
LB	Landing Barge
LBK	Landing Barge, Kitchen
LBV(M)	Landing Barge, Vehicle (Motorised)
LCA	Landing Craft, Assault
LCA(HR)	Landing Craft, Assault (Hedgerow)
LCB	Landing Craft, Barge
LCG	Landing Craft, Gun
LCH	Landing Craft, Headquarters
LCI	Landing Craft, Infantry
LCI(L)	Landing Craft, Infantry (Large)
LCI(M)	Landing Craft, Infantry (Medium)
LCM	Landing Craft, Mechanised
LCP	Landing Craft, Personnel
LCP(L)	Landing Craft, Personnel (Large)
LCP(R)	Landing Craft, Personnel (Ramped)
LCS(L)	Landing Craft, Support (Large)
LCS(M)	Landing Craft, Support (Medium)
LCT	Landing Craft, Tank
LCT(A)	Landing Craft, Tank (Armoured)
LCT(R)	Landing Craft, Tank (Rocket)
LCVP	Landing Craft, Vehicle and Personnel
LSD	Landing Ship, Dock
LSE(LC)	Landing Ship, Emergency Repair (Landing Craft)
LSH	Landing Ship, Headquarters
LSI	Landing Ship, Infantry
LSI(L)	Landing Ship, Infantry (Large)
LSI(M)	Landing Ship, Infantry (Medium)
LST	Landing Ship, Tank

ABOVE Three British-crewed GMC DUKWs under-way during the loading in England of Force L on 4 June. The 2½-ton amphibious DUKW truck was widely used by British and US forces and could carry a cargo of some three tons of equipment and stores. In the background can be seen two LST. *(IWM B5154)*

BELOW Landing Craft, Tank (LCT) and Landing Craft, Headquarters (LCH) vessels moored at Southampton ready to sail for Normandy. More than 800 LCT were used on D-Day, making them the most numerous of all Allied vessels used in the landings. *(IWM A23731)*

(Landing Craft, Personnel), LCA (Landing Craft, Assault), LSI(S), (M), and (L) (Landing Ship, Infantry (Small), (Medium) and (Large)) and LCI(S) (Landing Craft, Infantry (Small)). In the order of battle there were also the specialised support landing craft like the LCT(A) (Landing Craft, Tank (Armoured)), LCT(R) (Landing Craft, Tank (Rocket)), LCG (Landing Craft, Gun) and LCF (Landing Craft, Flak). The inner man was not overlooked in this roll-call of vessels because there were LBKs (Landing Barge, Kitchen) and LBBs (Landing Barge, Bakery), which cooked for the troops ashore.

ABOVE Landing Craft, Infantry (Large) 93, commanded by Lt Budd B. Bornhoft, USCGR, took part in the landings on Omaha beach on D-Day and suffered heavy damage from enemy gunfire and mines. Her hull was holed numerous times and she became stranded between the beach and an offshore sand bar. Under continuous enemy fire, she was abandoned. *(USNA)*

BELOW HMS *Glenearn*, Landing Ship, Infantry (Large). The role of the LSI(L) was to carry troops to within a few miles of the landing beaches where they would be transferred to LCA, lowered into the sea by davits, then taken to the landing beaches. The 9,748-ton *Glenearn* had been converted from a requisitioned merchant vessel and was one of a number of LSI(L) used in Normandy. *(IWM A25032)*

ABOVE Landing Craft, Mechanised (LCM). More than 8,500 LCMs were built in American yards between 1942 and 1945. Britain had no tank-carrying assault craft smaller than the LCT (Landing Craft, Tank), so LCM were readily adopted by the Royal Navy. LCMs could either be carried on ship davits, or be towed across the English Channel. Because of the danger of swamping in rough seas, most LCM were towed across the Channel empty, and were not loaded with troops until the morning of the landings. LCM can be easily identified by their unique perforated bow ramp. Some 464 LCM were assigned to Operation Neptune. *(USNA)*

BELOW British troops wade ashore from LCT beached in the shallows. Also in the picture is an LCM and, swimming ashore on the right, a Universal Carrier; at the extreme left can be seen the canvas flotation screen of a Sherman DD. *(USNA)*

Specialised support craft

H ere are some examples of the specialised support craft that took part in the Normandy landings. There were many more that fulfilled specific support roles on and after D-Day, but space does not permit inclusion of them all.

Amphibious truck – the six-wheel-drive DUKW

The six-wheel drive DUKW amphibious truck was the brainchild of the American General Motors Corporation (GMC) and a famous New York city yacht design firm, Sparkman & Stephens. It was the latter that bestowed on the DUKW its impressive sea-going capability. Equally at home on water as it was on land, the tough and versatile DUKW was widely used in Normandy for transporting goods and troops to the beaches from LCTs and other craft lying offshore. Some 21,147 DUKW were eventually built.

Contrary to some accounts DUKW is not an acronym. In fact it was manufacturer GMC's code for a type of military, wheeled amphibious landing craft: D = 1942, U = utility (amphibious), K = all-wheel drive, W = two powered rear axles.

Prior to D-Day, the DUKW was central to an experiment that was looking for a better way for US Rangers to scale the vertical cliffs on Omaha Beach at Pointe du Hoc. The US Army took inspiration from fire brigade turntable ladders, fitting a standard 100ft extendable fireman's ladder into the hold of a DUKW, with power for extension and elevation supplied from the vehicle's engine. Retractable jacks were fitted on each side of the vehicle to give added stability when the ladder was extended. Two automatic weapons could be mounted and fired from the top of the ladder by a gunner, who could be elevated with the ladder and was protected by armour plating.

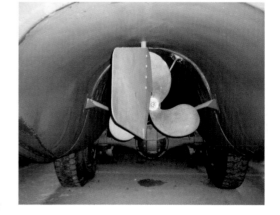

Three DUKWs were adapted to carry ladders and US Rangers trained with them on the cliffs at West Bay in Dorset, but it was found to be impossible to get a DUKW close enough to the foot of the cliffs to extend its ladder, so the scheme was dropped.

Landing Craft, Headquarters

Some 21 LCI(L) were converted for use as headquarters ships in the assault area for local senior officers. Extra accommodation was provided for six officers, eight petty officers and 48 ratings in addition to the vessel's operating crew. LCH were also equipped with offices, a sick bay and communications equipment.

Landing Craft, Assault (Hedgerow)

Fitted with spigot mortars – a lesser version of the Hedgehog mortars used by anti-submarine vessels – the LCA(HR) was expected to deal with anti-tank mines and barbed wire on the landing beaches above the high-water mark before the first troops landed. Due to the state of the tide it proved impossible for the mortars with a range of about 400yd to reach the beach minefields. The patterns instead fell on the beach obstructions.

Around 45 LCA(HR)s sailed for the British beaches, towed by LCT; all encountered difficulties on the passage across the Channel owing to the bad weather, with a significant number foundering, and in the end only about

12 craft made it to the beaches. Experience proved that the LCA hull was not suitable for a long sea passage in anything but fine weather, and it was not able to withstand the weight of mortar firing – some craft had their bottoms almost blown out.

Landing Craft, Tank (Armoured)

Some 69 LCT Mk 5 were converted to LCT(A) specification. This involved fitting armour plate over the engine room, fuel tanks and wheel house, and the construction of tank ramp platforms on the tank deck to carry two Centaur tanks of the Royal Marine Artillery, each mounting a 4in gun.

ABOVE A DUKW disembarks from an LST on to a Mulberry pier head pontoon. *(USNA)*

LEFT Landing Craft, Tank (Armoured). Armour plating was added to some LCTs in early 1944 in the UK prior to a reverse Lend-Lease to the US Navy, where the craft were redesignated LCT(A). The modification meant they could run in to the beach carrying two tanks side by side firing over the bow ramp. One of 48 such craft intended for use on D-Day, LCT(A) 2402 was damaged in the Normandy landings. *(USNA)*

It was intended that Centaurs of the Royal Marine Assault Support Regiments were to provide direct fire on to the beaches during the run-in, indirect fire from the water's edge as called for by FOOs, and thickening-up of artillery fire under command of field regiments to which they were attached. Their fire support was to be critical in the minutes between the landing of the first troops and the arrival of the SP artillery. On D-Day the Centaurs did not fire on the run-in and only half their number actually made it ashore.

Their contribution to the fire plan was seriously compromised by a number of factors: substandard workmanship at the shipyards where the conversions were carried out meant that the LCT(A) were unseaworthy and prone to flooding between the armour plating and decks (which were not watertight above the water line); overloading – the average load of an LCT(A) was 80 tons with the addition of 35 other officers and men; the Navy's inability to land the LCT(A) at the right time and place thanks largely to the bad weather, which caused a disproportionate number of casualties

and forced a large number of craft to turn back to the UK (with flooded engine rooms and in need of a tow), and others to founder – all of which seriously interfered with the ability of the Marine Artillery to fulfil its role on D-Day. On Gold Beach only two out of 16 LCT(A) allocated to Force G beached on time and took part in the fire plan. This was a story repeated on the other beaches.

Landing Craft, Gun (Large)

Navy destroyers seldom would have been able to get close enough inshore to bombard concrete beach defence positions, and would have been vulnerable to artillery fire from the shore, so the task was given to the Landing Craft, Gun (Large), which proved invaluable for this purpose and for close support on D-Day. The LCG(L) was converted from the LCT Mk 3 or Mk 4 to take a pair of ex-destroyer 4.7in Quick-Firing guns and 2 to 4 x 20mm Oerlikons, mounted on a false deck built over the hold. However, once the initial bombardment was over the LCG(L) were not used again in Normandy.

BELOW Landing Craft, Gun (Large). LCG(L) 939 was part of the Support Squadron Eastern Flank, attached to the British 4th Special Service Brigade in support of 41 and 48 RM Commando in the Juno sector.
(IWM A23752)

LEFT Landing Craft,
Tank (Rocket). Adapted
from the LCT Mk 2
and Mk 3, the LCT(R)
was designed for
shore bombardment,
showering the beaches
with up to 1,064 RP-3
60lb rocket projectiles
before the first troops
went ashore. It took the
19-man crew between
four and six hours to
reload the entire rocket
battery. Some 36 took
part in Neptune.
(IWM B5263)

Landing Craft, Tank (Rocket)

Of all the specialised gunfire support craft used
on D-Day, the 196ft-long LCT(R) was the most
important. The vessels were to provide close
support during the initial stages of an opposed
amphibious landing, raining HE fire on to an
area target on the beach prior to touchdown of
the first wave of assault landing craft. Two types
of LCT, the Mk 2 and Mk 3, were converted to
the LCT(R) specification to carry 792 and 1,064
x 5in rocket projectors respectively. Rocket
racks were attached to an upper deck at an
angle of 45°, with six projectors per rack. The
rocket racks were not adjustable so the only
way of aiming the rockets was by aiming the
vessel itself. The LCT(R) was capable of firing
between 800 and 1,000 29lb HE projectiles
electrically in salvoes of 24, 26, 39 or 42
rockets per salvo, from a range of 3,500yd, with
a complete reload for all projectors stowed in
magazines below the upper deck. The craft's
range from the beach was determined by
onboard radar and its position was fixed with a
Brown Gyro Compass.

LEFT For whoever
was on the receiving
end, an LCT(R)
in action was a
frightening sight and
sound. Each of its
rocket warheads had a
bursting radius of 30yd
and it was thought that
the shock power from
one salvo of rockets
was equal to two and
a half times that of a
battleship salvo. One
observer on D-Day
recalled the sound of
rockets being fired
'was like calico being
ripped apart'. (USNA)

Thames dumb barge conversions

Thames lighters or dumb (engineless) barges were requisitioned and converted into naval landing barges, some 400 of which (manned by 3,500 men) supported operations off the Normandy coast from June 1944.

As the LBV (Landing Barge, Vehicle) was fitted with engines, rudder and a ramp, their primary task was to carry vehicles and supplies from ship to shore, particularly equipment that was too big, bulky or heavy for DUKW – a task at which the LBV excelled.

The second major use of the converted dumb barges was to form Supply & Repair (S&R) flotillas that provided fuel, water, hot meals, repairs and maintenance to the many hundreds of landing craft on both the American and British beaches.

These specialised vessels included LBO (Landing Barge, Oiler), LBW (Water), LBK (Kitchen), and LBE (Emergency Repair). Finally,

ABOVE Landing Barge, Kitchen (LBK). Crews of small landing craft engaged in landing troops and supplies on the invasion coast of France line up on a LBK for a hot midday meal on 16 June 1944, while an LCVP and an LCM await their turn to come alongside. Affectionately called 'Micky's Fish and Chip Bar', the floating kitchen was a converted Thames barge and served up more than 1,000 hot meals on D-Day. Ten LBKs provided hot meals for the men in hundreds of small craft that ferried supplies to the Normandy beaches from ships anchored offshore. They had a complement of 25 men – 13 cooks, nine seamen and three stokers. Every day and in all weathers hot meals for between 500 to 700 men were served. *(IWM A24015)*

but of more limited use, were LBF (Landing Barge, Flak) and LBG (Landing Barge, Gun). On D-Day and afterwards on the Normandy coast there were 10 Supply & Repair Flotillas, each of which comprised six LBE, 10 LBO, two LBW and one LBK.

LST conversions

The ferrying of locomotives after D-Day had been taken care of with the five train ferries (see caption), but the ferrying of wagons presented an altogether different problem. With cargo space on merchant ships at a premium, the idea of adapting LST to carry rolling stock was taken up, and eventually 52 were modified for the purpose, with most of the vessels converted before D-Day. The essential requirement was that the installation of rail tracks and fittings on the tank decks of the LST should not interfere in any way with the operation of the vessel as a carrier of tanks, tracked vehicles and motor transport. Rail track was tack-welded to the deck and a concrete slope was made on the inboard side of the bow doors to allow wagons to pass over the 'hump'. Fabricated three-way stools with link boxes installed in the bows of the LST enabled the links from the ramp runway to be dropped into the link box of the track being loaded. The stools remained in place on the vessel so long as rolling stock ferrying was being undertaken, but were easily removed if road vehicles had to be loaded.

ABOVE Perhaps the most unusual vessels to be used in the wake of the Normandy landings were the converted British train ferries for carrying locomotives to the Continent. Hampton Ferry was one of three ex-Southern Railway 'Twickenham'-class train ferries pressed into service for the task after D-Day. Before the war they had operated mainly between Dover and the French Channel ports. Two LNER 'Daffodil'-type ferries that had plied the crossing between Harwich and Zeebrugge were also used. Hampton Ferry was 346ft 8in long with a beam of 60ft 7in and a gross registered tonnage of 2,839 tons. (USNA)

LEFT Railway wagons are offloaded from LST 21 at a French port. (USNA)

LCT(A) and the run-in shoot

An important element of the Allied plan to suppress enemy beach defences on the morning of D-Day was the run-in shoot on the British and Canadian beaches by divisional artillery of the 3rd British, 50th (Northumbrian) and 3rd Canadian Divisions, carried in LCT and LCT(A).

It was an integral part of the overall joint fire plan for the invasion provided by the Royal Navy (naval bombardment) and the RAF (air bombardment), which included battleships, cruisers, destroyers and medium bombers in their plan. Three self-propelled (SP) artillery regiments were allocated to each of the two British divisions, while four SP regiments supported the 3rd Canadian Division.

Specially adapted LCT(A) carrying SP guns and Centaur tanks (with 95mm guns, acting as SP artillery) were to fire at onshore targets as they approached the beaches. These targets included fortified positions, *wiederstandneste*, machine-gun emplacements and observation posts. Each of the artillery regiments supporting the 3rd British and 3rd Canadian Divisions was equipped with 24 105mm Priest self-propelled (SP) guns, while those with the 50th Division used the 25pdr Sexton SP gun (also known as the Ram) – a total of 240 SP guns.

Gun ranging was initiated at about 10,000yd out from the shoreline and firing commenced at about 9,000yd, continuing for 30 minutes during which time the SP guns kept up a steady barrage on to the beaches above the heads of the infantry approaching the shore in their LCA. The SP fire was directed by a Forward Observation Officer (FOO) travelling in a small assault craft ahead of the LCT, who was in direct contact with them by wireless. The LCT maintained their correct line of approach by wireless signals from accompanying motor launches fitted with Outfit QH and Type 970 radar. During the run-in, the 10 regiments of SP artillery fired 1,800 rounds each in about 30 minutes.

Individual craft eventually broke away as close as 1,500yd from the shoreline as LCA carrying the assault troops touched down on the beaches. The LCT carrying the SP artillery were beached later, between one and two hours after H-Hour (8:25am and 9:25am). Then the various SP field regiments went into action on the narrow strip of beach, their fire being directed on to targets by FOOs.

No individual element of the joint fire plan could be said to have stood out, but despite the rough seas and poor visibility over the beaches the SP artillery contributed to the overall neutralisation of enemy beach defences, as well as giving moral support to the infantry in the LCA heading for the beaches.

When comparing the experience of the SP artillery to that of the Royal Marine Artillery and its Centaur tanks, the failings of the LCT(A) are underlined. They were overloaded on D-Day, which made them top-heavy, and were not seaworthy thanks to the mad rush in the dockyards to convert them, which resulted in substandard workmanship. Many of the craft carrying Centaurs either sank or broke down in mid-Channel, turned back to England with technical trouble, or when they finally made it to the beach the Centaurs nosed off the vehicle ramps and drowned in deep water, which meant that only half of their number eventually made it ashore.

The US forces on Omaha and Utah were also supported by a run-in shoot with M7 Priest 105mm SP artillery of the 58th, 62nd and 65th Armoured Field Artillery Battalions, and by standard M4 Sherman gun tanks of the 70th, 741st and 743rd Tank Battalions in LCT and LCT(A). Some 18 LCT(A) were used on Omaha and eight on Utah for the run-in shoot, but the rough sea conditions made accurate fire difficult so in the end their part in the fire plan was small.

Assault craft that won D-Day

If there are three assault vessels that define the Normandy landings and whose invention changed military thinking, then they are the Landing Ship, Tank (LST), the Landing Craft, Assault (LCA) and the Landing Craft, Vehicle and Personnel (LCVP).

Landing Ship, Tank (LST)

The 328ft-long, 4,080-ton LST was a clever compromise between a conventional cargo ship and a landing craft. Its forward end was flat-bottomed to enable it to be run aground on or close to the beach, while its high-peaked bow was fitted with a pair of clamshell doors that opened to let down a ramp connected to the tank deck for unloading internal cargo. Once unloaded, the LST would either wait until the next high tide came and floated it off the beach, or the vessel would winch itself off using anchors previously dropped astern. The plan for D-Day was to lower the LST ramps offshore and disgorge cargo on to LCT and Rhino ferries which would then carry it to the shore. Although slow and hard to handle at sea, the LST (nicknamed Large Slow Target for obvious reasons) was a vital link in the initial assault and subsequent resupply operations.

ABOVE Landing Ship, Tank (LST) 543 comes alongside an LST pier head in Mulberry A. Each bow door of the LST is 24ft high and 14ft 11in wide. They are operated by two 3hp bow door drive units, each consisting of a gear motor driving a screw through open gearing. *(USNA)*

LEFT Rhino ferry. A flat-top pontoon powered by a pair of 143hp outboard motors, designed by the US Navy Civil Engineering Corps to bridge the gap between ship and shore and used for ferrying supplies and equipment. This loaded Rhino ferry was one of 72 used on D-Day and is seen from the loading ramp of an LCT between Saint-Laurent and Vierville. *(IWM A23999)*

ABOVE Displaying the LST's beaching capability, Royal Navy LST 406 is seen on the Normandy shore with her bow doors open and ramp down, while a 22nd Armoured Brigade HQ, 7th Armoured Division, Crusader AA Mk III tank is driven ashore on 7 June. The bow ramp is 23ft long and 16ft 3in wide, while the actual opening into the tank deck is 13ft high and 15ft wide. *(IWM B5129)*

RIGHT The cavernous tank deck of the LST was frequently used for transporting casualties back to England. They were modified to carry up to 300 stretcher cases. In this photograph, British and Canadian wounded can be seen in the racks on either side of the deck; those men on the deck are wounded Germans. *(IWM A25102)*

Landing Craft, Assault (LCA)

Designed by Ken Barnaby and built in Britain by John I. Thornycroft Ltd, the LCA was widely used by British and Commonwealth forces throughout the Second World War. Like its American counterpart the LCVP, its primary purpose was to ferry troops from offshore troop transports to land them on enemy-held shores. Made from hardwood planking and clad with light armour plate (¾in on bulkheads and sides, and ¼in on decks) and powered by two 65hp Ford V8 petrol engines, this 41½ft-long shallow-draught boat with bow ramp had a crew of four and could carry 36 fully equipped infantrymen, or 800lb of cargo, at a speed of seven knots. It was armed with a Bren light machine gun or two Lewis guns. The first orders were placed in 1939, and by the end of 1944 some 1,694 LCA had been built by Thornycroft and its subcontractors in small boatyards and by the furniture industry. On D-Day, Royal Navy LCA put troops ashore on Juno, Gold and Sword Beaches in the British and Canadian sectors, and landed formations of US infantry on Omaha Beach, as well as the Rangers who made the assault on the Pointe du Hoc. LCA were also used in the westernmost landings on Utah Beach and in the pre-dawn landing on the Îles Saint-Marcouf.

BELOW Foxtrot 8 is an LCA that was built in the late 1950s by Camper & Nicholson at Gosport. She is very similar in design to those LCA used in the D-Day landings. In 1982 Foxtrot 8 saw active service in the Falklands campaign with HMS *Fearless*, but is now in a sorry state of repair and the subject of a possible future restoration by the Portsmouth Historic Boats Trust. A sister craft to Foxtrot 8 can be seen at the Royal Marine Museum, Southsea. It is believed that these two craft are the last remaining of their type. *(Author)*

LEFT Landing Craft, Assault vessels moored in Weymouth harbour before D-Day, with American troops and Royal Navy crew. In the background are Landing Ship, Infantry (Large) assault vessels. LCI(L)-87 on the left was flagship to Commander Miles Imlay, commander of USCG Flotilla 10. As Imlay's flagship, who was concurrently serving as a deputy assault commander, LCI(L)-87 took part in the landings at Omaha beach on 6 June and remained off the beach all day, under constant enemy fire, as Imlay directed incoming vessels to their assigned landing areas. *(USNA)*

Landing Craft, Vehicle and Personnel (LCVP)

The LCVP – the 'VP' or 'Higgins boat' – was probably the most important US landing craft of the Second World War. Introduced to operations in the autumn of 1942, it carried the most American and many Allied troops to landing beaches. Numerous photographs taken on D-Day of troops leaving landing craft are of LCVPs, and it is this famous assault craft that we will now look at in more detail.

The 'Higgins boats'

'Andrew Higgins,' said US President Eisenhower in 1964, 'is the man who won the war for us.' With the help of a team of American subcontractors, Higgins Industries in New Orleans, Louisiana, built more than 20,000 landing craft that delivered tens of thousands of American and Allied troops on to North African and European shores, and beaches in the Pacific, during the Second World War.

It was these flat-bottom landing craft – particularly the LCVP and the LCM (Landing Craft, Mechanised) – that made the D-Day landings in Normandy possible, not to mention those at Guadalcanal, Tarawa, Iwo Jima, Okinawa, Leyte and Guam, and hundreds of other lesser-known assaults. In fact, the 'Higgins boats' landed more Allied troops during the war than all other types of landing craft combined. Eisenhower went on

to explain: 'If Higgins had not designed and built those LCVPs, we never could have landed over an open beach. The whole strategy of the war would have been different.'

The founder and president of this boat-building company in the southern states of the USA was Andrew Jackson Higgins (1886–1952), a quick-tempered outspoken Irishman, who was blessed with the ability and vision to turn ideas into reality. His rise to international prominence during the Second World War was through his design and mass production of small combat boats for the US Navy. Thanks to Higgins, the Allies no longer needed to sweep harbours of mines and take over enemy-held ports before they could land an assault force. 'Higgins boats' conferred on them the ability to transport thousands of men and hundreds of tons of equipment swiftly through the surf to less-fortified beaches, bypassing established harbours. This new amphibious capability was to change the strategy of modern warfare forever.

The genesis of the 'VP' lay in the wetlands of south Louisiana during the 1930s when Higgins Industries developed a workboat, nicknamed the 'Eureka' model, which was designed for work in the swamps and marshes. With its shallow draught, not only could the boat operate in just 18in of water, run through vegetation and over logs and debris without fouling its propeller, but it could also run right up on to the shore and extract itself without damage.

The strongest part of the boat was the 'headlog' – a solid block of pine at the bow that enabled the craft to run at full speed over floating obstacles and sandbars, and right up on to the beach without damaging its hull.

A deep V-shaped hull forward led to a reverse-curve section amidships and two flat planing sections aft, which flanked a semi-tunnel beneath the hull that protected the propeller and shaft. When the boat was moving, aerated water flowing under the forefoot of the craft created less friction, allowing it to achieve faster speeds and greater manoeuvrability. The reverse curve feature meant that objects in the water would be pushed away from the boat at a point between the bow and amidships, including the aerated water (only solid water reached the propeller). This enabled continuous high-speed running and minimised damage to the propeller, because floating objects seldom came near it. The flat planing sections aft, on either side of the shaft tunnel, had a catamaran effect that added to the hull speed. Combined, these features enabled the successful adaptation of the 'Eureka' boat to a landing craft, and when a bow ramp was fitted at the request of the US Marine Corps, the war-winning LCVP design was complete.

By the time war broke out in Europe, Higgins was producing workboats and prototype landing craft in a small warehouse behind his New Orleans showroom. When the US government began placing orders in 1941 for his craft for Navy and Marine Corps use, construction of Higgins boats expanded rapidly. At the peak of production the combined output of Higgins Industries, working 24 hours a day, exceeded 700 boats a month.

Essentially there were two classes of military craft designed and produced by Higgins – high-speed PT boats and various types of Higgins landing craft (LCP, LCPL, LCVP, LCM). The latter were built from wood and steel and were used for transporting fully armed combat troops, light tanks, field artillery, equipment and supplies that were essential to amphibious operations.

The 'VP' boat could land a platoon of 36 fully equipped infantrymen, or a Jeep and 12 men, on a beach – then extract itself quickly, turn

Higgins Industries – the first equal opportunity employer

Higgins was ahead of his time when it came to employment policy. His company was the first equal opportunity employer in New Orleans with a racially integrated workforce, employing a wide range of different individuals, including women, men who had not been drafted into the forces, the elderly, blacks, and handicapped people. Regardless of their race, sex, age or physical disability, all were paid equal wages according to their job rating. Along with employee housing and childcare facilities, Higgins looked after the welfare of his 30,000-strong workforce, and they in turn went the extra mile for the company to produce a world-class product that helped win the war.

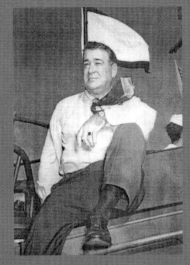

ABOVE Andrew Higgins, designer and builder of the LCVP – the 'Higgins boat'. *(Copyright unknown)*

around without broaching in the surf, before heading back to the landing ship to pick up more troops and/or supplies.

Constructed of durable veneer marine mahogany plywood, the 'VP' was 36ft 3in long and 10ft 10in in the beam, and included a solid pine log in the bow that acted as a buffer against any collision with objects in the water, thereby preventing hull damage. The craft was defended by two .30-calibre machine guns and was fitted with a steel bow ramp that protected its human cargo (or 8,000lb of supplies and equipment) from gunfire, as well as facilitating their swift disembarkation on the beachhead. The 225hp Gray marine diesel engine, which gave the 'VP' a top speed of 12 knots, was controlled by the coxswain using a simple throttle/gear shifter. Two rudders, one for normal control, the other forward of the propeller for backing off, allowed the craft to quickly power on and back off beaches.

ABOVE US Rangers board an LCVP from the deck of the assault transport USS *Samuel L. Chase* (APA-26). An item of equipment instantly recognisable as 'Ranger kit' was the assault vest (seen here). Issued to officers and NCOs, the vest provided a lot more carrying capacity than other equipment-carrying webbing issued to US assault troops. It was from *Samuel L. Chase* that the famous war photographer Robert Capa embarked on a landing craft for Omaha beach at 4:30am on 6 June, where he took some of the defining photographs of **D-Day.** *(US Coast Guard Collection/ US National Archives – USCGC/USNA)*

LEFT Troops crouch inside an LCVP on their run-in towards the shore at Omaha. *(USCGC/USNA)*

ABOVE The time is 7:00am on 6 June. In this image, LCVP PA26-15 from *Samuel L. Chase* approaches the Easy Red–Fox Green beach sector on Omaha carrying troops of 1/16th Regimental Combat Team (RCT), billowing white smoke as a result of taking a hit that set off a grenade on board. After unloading troops, the boat's coxswain, Delba L. Nivens, assisted by two other crewmembers, put out the fire and returned the landing craft to the *Samuel L. Chase*. On the beach can be seen the many obstacles placed there by the Germans – hedgehogs, ramps and pilings; and GIs who are sheltering on the shingle bank. *(USCGC/USNA)*

RIGHT A pair of LCVP from *Samuel L. Chase* plough through the surf to put their troops ashore on Omaha. *(USCGC/USNA)*

LCVP in action on D-Day – landing on the beaches

BELOW A US Coast Guard-manned LCVP from *Samuel L. Chase* disembarks troops of E Company, 16th Infantry, US 1st Infantry Division on Omaha on the morning of 6 June. *(USCGC/USNA)*

RIGHT A cartoon from the US Navy's LCVP official handbook. Humour was an effective way of putting across serious messages. *(US Navy)*

RIGHT Soon to be flung into the maelstrom of action on Omaha, the expressions on the faces of these men range from contemplation to apprehension. *(USCGC/USNA)*

BELOW LCVP from the attack transport *Joseph T. Dickman* (APA-13) land troops of the US 4th Infantry Division on Utah Beach. *(USCGC/USNA)*

Chapter Three

Corncobs and Whales

The Mulberry harbours

'I cannot finish my task without expressing in a single, simple sentence the admiration which I have felt, as a civilian wholly unconnected with this enterprise [the Mulberry harbours], for the imagination, resource, resolution and courage of those who planned and carried it out.'

(Sir Walter Monckton in his Report to the Chiefs of Staff Committee, 18 January 1946)

OPPOSITE The beach at Arromanches today is a scene of calm, where at low tide tourists can walk among the washed-up remains of the Mulberry harbour. *(iStock/Mikael Damkier)*

'Mulberry' was the codename given to the artificial harbours assembled off the coast of Normandy in June 1944. Mulberry A (American) was built off Omaha Beach at Saint-Laurent-sur-Mer and was for use by the American invasion forces. Mulberry B (British) was assembled off Gold Beach at Arromanches for use by the British and Canadian invasion forces. Both harbours were of a similar size to the Channel port of Dover and comprised all the elements one would expect of any regular harbour.

Mulberry was also the collective name for the sum of the parts of these artificial harbours, created from a variety of different structures, comprising a floating outer breakwater (Bombardon), a static breakwater (Corncob) and reinforced concrete caissons (Phoenix), floating piers and roadways (Beetle and Whale) and pier heads (Spud).

Without an effective means of supplying the Allied invasion armies in Normandy, the plans to liberate Europe were certain to falter and fail. The disastrous Dieppe raid of 19 August 1942 confirmed what most Allied planners already knew – that an attempt to penetrate the Atlantic Wall and capture a well-defended port on the north French coast was unlikely to succeed without heavy and effective naval and air support. The major ports of Cherbourg and Le Havre at either end of the proposed invasion coast were heavily defended and were likely to remain in German hands for some time after the initial assault. Indeed, a full-scale attack from the sea to seize Cherbourg would have been a long and bloody business, with little freedom of manoeuvre, and the enemy would have had a relatively easy task of containing the Allied force in the Cotentin peninsula if there were no simultaneous landings on the

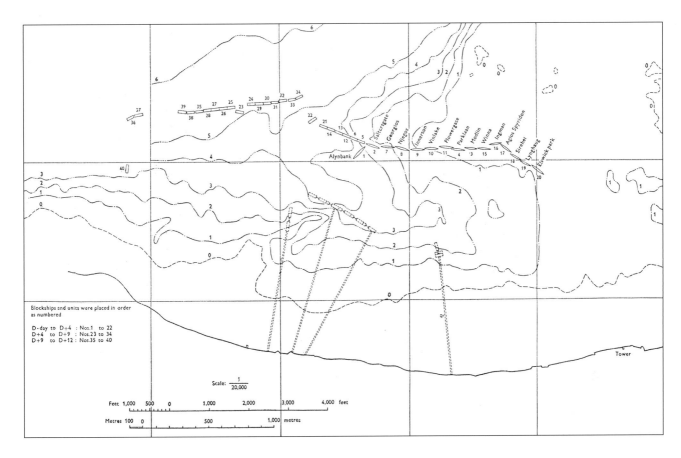

Blockships and units were placed in order
as numbered

D-day to D+4 : Nos.1 to 22
D+4 to D+9 : Nos.23 to 34
D+9 to D+12 : Nos.35 to 40

Scale: 1/20,000

Feet 1,000 500 0   1,000   2,000   3,000   4,000 feet
Metres 100 0   500   1,000 metres

Tower

Calvados beaches. Nor was the alternative of landing supplies over open beaches any more acceptable owing to the vagaries of the weather, which could easily bring a halt to operations. It was for these reasons that Allied planners identified the need for one or more artificial harbours to fulfil the logistical needs of the invading armies.

The story goes that at a meeting following the Dieppe raid, Vice-Admiral John Hughes-Hallett (who had been the naval commander for the raid) declared that if a port could not be captured, then one should be taken across the Channel. His suggestion was greeted with mild derision from many who were present, but Hughes-Hallett had the support of a powerful ally in the Prime Minister, Winston Churchill. When Hughes-Hallett was later appointed Naval Chief of Staff to the Operation Overlord planners, the concept of what became known as Mulberry harbours began to take form.

In the autumn of 1942 Vice-Admiral Lord Louis Mountbatten, Chief of Combined Operations, outlined the specifications for a pier and pier head, which called for piers at least a mile long and able to carry a continuous flow of traffic, and a pier head capable of handling three 2,000-ton coasters. Three proposals were tendered for consideration, and through the spring and summer of 1943 designs

for breakwaters, piers and pier heads were considered and trialled to assess their suitability.

The first was by Iorys Hughes, a consulting engineer to the War Office, who had designed the Empire swimming pool at Wembley. His concept was for a fixed pier and pier head comprising a huge concrete caisson (called a 'Hippo') and a tubular steel bridge span (a 'Croc').

The second was from Ronald Hamilton at the Department of Miscellaneous Weapons Development (DMWD). Nicknamed the 'Swiss Roll' it was a floating roadway made from wooden planks strung together with steel cables. Under test it proved unable to carry loads greater than seven tons, which effectively ruled it out (although a version was incorporated as one component of the Arromanches Mulberry).

The third design was by the War Office department codenamed 'Transportation 5' (Tn5) under Major General D.J. McMullen, which had responsibility for port engineering and maintenance. Their proposal was for a floating bridge linked to a pier head, the latter fitted with adjustable spud legs that could be raised or lowered depending on the tide.

All three competing designs were built and tested in Scotland at Rigg Bay on the Solway

Firth, but it was Tn5's design that made the grade and was chosen. Designed by Lieutenant-Colonel William T. Everall and Major Allan Beckett, their scheme was adopted and manufactured under the management of J.D. Burnal and Brigadier Bruce White, the Director of Ports and Inland Water Transport at the War Office.

By the end of July 1943 the planning for Overlord was well advanced and the need for an artificial harbour had been acknowledged. It was also accepted that such a harbour would need to be prefabricated in the UK and towed across the Channel, since neither the facilities nor the time for major construction work would be available in Normandy immediately after the invasion.

At the Quebec Conference in Canada during August 1943, the plans for Overlord were debated and accepted by Churchill, Roosevelt and the Combined Chiefs of Staff, and the proposal for separate British and American artificial harbours was also approved. Two Mulberry harbours were to be created – Mulberry A off Omaha Beach in the American sector, and Mulberry B at Arromanches in the British and Canadian sectors – to provide a refuge for large numbers of small landing craft and sheltered lightering anchorages for stores ships, including Liberty ships (mass-produced cargo ships built in the USA between 1941 and 1945). They would also provide the necessary port facilities to offload the thousands of men, vehicles and equipment, as well as tons of supplies ranging from ammunition to rubber tyres, all necessary to sustain Operation Overlord and the ensuing Battle of Normandy.

After the war Brigadier Bruce White was often asked why the name 'Mulberry' was chosen for the harbours. This is his explanation:

The simple answer is that following my return from America [Washington, after the Quebec Conference] I found on my table at the War Office a letter that had no security and was headed 'Artificial Harbours'. I felt this was a possibility where security might be broken and therefore approached the head of the security branch at the War Office and requested a meeting at which I explained the case, and the chairman asked me what I would like to be done. I said utilisation of a codeword would give me the security I would wish for. Turning to the junior officer behind him, he asked what was the next name from the code book, and the officer replied, 'Mulberry'.

The Combined Chiefs of Staff reported on 2 September that the Overlord plan required a minimum discharge through the two harbours of 12,000 tons per day (5,000 tons in the American and 7,000 tons in the British harbour), exclusive of motor transport (MT) and in all weathers. Facilities also had to be provided for the landing of MT and other stores from Rhino ferries, DUKW, pontoons and LCT on a protected beach. The landing of stores was to be partly from Spud piers and partly from ships moored in the harbour. On 4 September the signal 'Bigot, Most Secret' was received in London with orders to get to work with all haste on the two artificial harbours.

A short but intensive period of collaboration followed the conference between British and American engineers in which the basic design of the harbours was agreed upon. So too was the requirement for them to be fully operational within a fortnight of D-Day and to remain in use for 90 days.

However, infighting between the Admiralty and the War Office over who was responsible for what eventually led to the intervention of the Vice-Chiefs of Staff on 15 December 1943 to settle matters. They decided that the Admiralty would be responsible for the design of the Bombardons, the procurement of the blockships, for assembling all parts of the harbour on the south coast of England, and for towing all of the Mulberry components across the Channel, as well as for surveying, siting and layout of the harbour and breakwaters, and for marking navigational channels and moorings.

The War Office would be responsible for the design and construction of the Phoenix caissons and Whales, for their protection against enemy air attack and for 'parking' them until they were required for towing to Normandy. Once across the Channel the Army was also responsible for sinking the caissons in position as well as for the assembly of all the Mulberry harbour port facilities, and for its defence from static sites inland and weapons emplaced on the Phoenixes.

The Americans were committed to installing their own harbour (Mulberry A) at Saint-Laurent, from the British-built components. For this task they used the naval Corps of Civil Engineers, with installation work carried out by Force 128 (also known as Force Mulberry) using personnel drawn from 108 Construction Battalion – the famous 'Seebees'. The battalion was responsible for crewing the caissons and the Whale tows on their passage across the Channel, manning the Rhino ferries and pontoon causeways, as well as for the installation, operation and maintenance of all harbour equipment once across the Channel.

The site location for each harbour was determined by the purely military requirements of the assault, the Front of which was to cover the stretch of coastline from the base of the Cherbourg peninsula in the west to the mouth of the River Orne in the east (see page 11). The harbours had to be sited so that their construction did not interfere with the assault, and at a point where shore exits either existed or could be easily provided.

Surveying the sites

Vital information concerning the contours of the Normandy shore and seabed was obtained from pre-war Z (M) charts created by the UK's Hydrographic Department, aerial photographs and daring surveys conducted by a special team of hydrographers. In October 1943, the 712th Survey Flotilla was formed at HMS *Tormentor* at Warsash on the Solent with the purpose of gathering soundings off the enemy shoreline. Using LCP(L) (Landing Craft Personnel (Large)) with Royal Navy crews and a small contingent of hydrographers, they made six trips to the Normandy coast between November 1943 and January 1944.

To equip the LCP(L) for survey work it was fitted with an underwater exhaust to silence the engine, an echo sounder, taut wire gear and Outfit QH (Gee). The central deck well was covered with a weatherproof tarpaulin to make a blacked-out work area for the surveyors.

The first survey sortie, codenamed Operation KJF, was made in the moonless period (for purposes of concealment) on 26–27 November, when three LCPs obtained seven lines of soundings off Arromanches, where Mulberry B

Taking soundings

The method adopted by the 712th Survey Flotilla drew upon experience gained in 1942. To save fuel, and make for a quicker crossing, the designated LCP would sail from Cowes under tow by Motor Gun Boats (MGB). At a point some 30 miles off the French coast they would slip their tow and proceed silently inshore, while the MGB established a patrol line around the cast-off point. The LCP would use QH fixes (see page 33) until they were able to confirm their position by sounding for a specific depth contour (usually taken from the relevant Z(M) chart) and cross-referencing this with a visual sighting of a landmark. Here, one LCP would anchor as a mark boat, while the other vessel(s) would run out a series of sounding lines at tangents. If possible, the shore end of the sounding line would also be fixed with a visual sight of a known landmark. If a fix was not possible, then a compass course would be used instead. Working backwards, where the shore ends were fixed, the taut wire distances could be used to help establish the location of the mark boat.

While the sounding boats were at work, the mark boat would deploy a standard pole logship to record the direction and strength of the current. Each boat also carried a sounding lead loaded with tallow to obtain samples of the sea floor. At a prearranged time, no matter how much data had been obtained, the LCP departed the French coast to re-rendezvous with the MGB for their tow back to base.

An additional LCP was received by the flotilla in December, which allowed them to modify the sounding method. Instead of anchoring the mark boat and then running soundings in a star pattern around it, the new procedure saw the mark boat anchoring roughly 1½ miles off shore from which the sounding boats first ran a line parallel to the coast. Then, at a set distance, they turned through 90° to run a sounding line into the beach. If a number of such lines were run, then a set of parallel lines to the beach could be obtained.

BELOW Royal Navy Coastal Forces Motor Gun Boats (MGB) 502 and 503. *(USNA)*

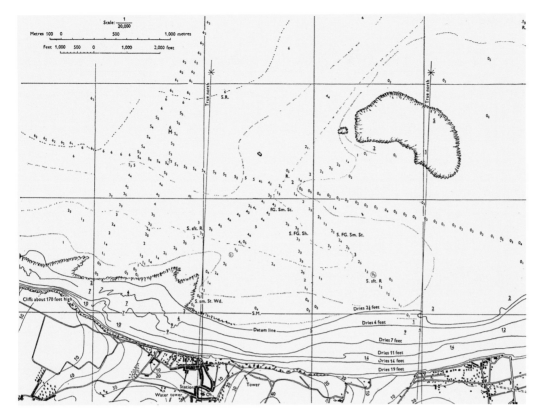

was to be sited. The same team attempted a similar feat on 1–2 December (Operation KJG) 12 miles to the west at the intended location for Mulberry A. However, owing to an error in the identification of a coastal landmark the soundings were wrongly made some 2,250yd east of where they should have been. The result was an incomplete set of soundings for the Mulberry A site.

The next moonless period was towards the end of December, when four LCP escorted by four MGB surveyed the coastline off the Point de Ver on the night of Christmas Day/Boxing Day (Operation Bellpush Able), but problems with the taut wire gear meant that only a single line of soundings could be taken, and the operation was cut short with little to show for their efforts.

Three days later, on the night of 28–29 December, the flotilla returned to the same area to complete the task, and this time they were successful. A moment of concern came when one of the LCP was heard by enemy lookouts and a star-shell was fired to illuminate the boat. Luckily the craft made good its escape into the darkness.

The flotilla's next sortie was two days later, but it involved no soundings because this time

the D-Day planners wanted information about the composition of the beach material and its sub-strata. On New Year's Eve 1943, two LCP set out to deliver a major and a sergeant of the Royal Engineers to the beach close to Point de Ver (see pages 20–21).

It was during the night of 30–31 January 1944 that the final pre-D-Day survey (codenamed Bellpush Charlie) took place after the planners had decided they needed more information on the potential site for Mulberry B. Three MGBs towed three LCPs from Cowes to a point about 30 miles off the French coast. Approaching the coast under their own power, the LCP were enveloped in thick fog, causing them to lose visual contact with one another. Even so, two of the boats were able to gather one line of soundings each, but when these were plotted later it was discovered they had both covered the same area. To make matters worse, the LCP were heard once again by the enemy, resulting in several star shells being fired, but the fog prevented them from being detected. With D-Day drawing nearer and the nights growing shorter, combined with the fact that the flotilla had been heard on two occasions, it was decided to abandon any further sorties.

The harbours
and their localities

BELOW Mulberry A,
the American Mulberry
off Omaha Beach on
15 June, showing the
Corncobs and the
Gooseberry harbour
area of calm water
sheltering a variety of
small craft in the lee of
the scuttled vessels.
(USNA)

The physical features of the two sites chosen for the Mulberry harbours in the Overlord plan were as follows:

Mulberry A: Located on an open sandy beach with a gradient of 1 in 100 below tide marks. The tidal range was 20ft at spring tides, and there was a tidal stream of up to 2.8 knots setting along shore.

Mulberry B: Located at the end of a beach enclosed by low rocks to the west, and protected by a bar and reef two miles out to the north-east. The tidal range was 21ft at spring tides, and there was a tidal stream of 2.3 knots setting along shore.

Both harbours were exposed to light winds from the north-west and east, and gales of Force 7 (35mph) were expected to be very rare during the 90-day period considered for the harbours. The lie of the coast provided protection against ocean swell, and waves greater than 100ft in length were thought most unlikely.

A massive undertaking

Building the components for the Mulberry harbours was a massive feat of civil engineering that took precedence over all other work in the British construction industry. The contracts were put out to commercial construction firms that included big names like Balfour Beatty, Costain, and Sir Robert McAlpine. Between December 1943 and when the first units were towed to France in June 1944, more than 55,000 men were employed on the construction of the Mulberry harbours at various locations around Britain.

Some 300 companies were involved in the construction of the pier heads to which ships would be moored for unloading, and some 250 more were building the floating roadways and pontoons to carry the vehicles on to the beaches. To ensure that absolute security was maintained, contractors were given only enough information to facilitate them in their immediate task of construction, and not one of them had drawings of the complete scheme.

Several factors complicated the construction and eventual deployment of components for the Mulberry harbours: a severe shortage of labour, particularly in skilled trades such as welders, steel fixers and riveters, electricians, carpenters and scaffolders; an equally severe shortage of materials, like steel, which were needed to build so many of the harbour's constituent parts; the existing heavy demands on dry docks for shipbuilding and repair meant there was a shortage of suitable dock facilities for the building of caissons and pier heads. Finally, there were not enough tugboats to assist the building programme before D-Day, and afterwards to tow the various components across the Channel to the Normandy beachhead.

When completed, the different Mulberry components were towed to their respective assembly points along the south coast – the Bombardons to Portland, blockships at Poole, the Phoenixes and Whales at locations between Southampton, Portsmouth, Selsey and Dungeness.

Mulberry harbour components

Below are details of the major elements of the Mulberry harbours.

Corncob and Gooseberry

In the event of bad weather the Mulberries would not have sufficient space to shelter anything up to 4,000 small craft as well as the hundreds of vessels that already used the two harbours. The solution lay in the provision of five areas of sheltered water, codenamed Gooseberries: two were included in the designs for the Mulberries and three subsidiary refuges composed entirely of blockships, one on the

BELOW Gooseberry No 2, Omaha Beach. Nearest the camera are Corncobs USS *George Childs* (577) and USS *Artemus Ward* (578). (USNA)

ABOVE The Dutch light cruiser *Sumatra* is steamed into position ready for scuttling. *(USNA)*

extreme western flank of the assault at Varreville and the other two east of Arromanches at Courseulles and Ouistreham. These were created by sinking a string of blockships (codenamed Corncobs). Once in position the Corncobs created Gooseberries.

The Chiefs of Staff gave their consent for 70 redundant merchantmen, as well as some old warships, to be sailed under their own steam (some had to be towed) to a location just off the invasion coast and sunk in a line to form a breakwater. To conform to the tidal conditions and the water depth at low water, the ships could only be sunk in about 12 to 15ft of water.

In stormy weather, shallow-draught vessels could shelter in the lee of the Corncobs, which also provided administrative facilities for first aid, repairs and refuelling. The superstructures

of the blockships provided accommodation for crews of the assault craft.

At a time when shipping was absolutely vital to Britain's war effort, it was stipulated that likely candidates for blockships should be about 30 years old, when it was considered their useful life was behind them. Indeed, about one-third of them had been, or were about to be, laid up, while the remainder were slow, inefficient and in constant need of repair. Suitable merchant ships were sought out and taken to Rosyth, Methil and Oban in Scotland for conversion. Four redundant warships were also selected to join the merchantmen – the First World War-vintage 25,500-ton 'King George V'-class super-dreadnought HMS *Centurion*, the 23,100-ton French dreadnought *Courbet*, the British 'D'-class light cruiser *Durban* (4,850 tons), and

Corncob crewman
Second Officer Lester Everett, Merchant Navy, SS *Empire Moorhen*

Merchant seaman Lester Everett was the navigating and gunnery officer on the SS Empire Moorhen, *an American hog-backed cargo vessel of First World War vintage. By 1944 she had completed her life as a cargo-carrying ship and was chosen as a Corncob for the final adventure of her long career. On either side of her bridge she displayed the identification number '307' in very large figures.*

SS *Empire Moorhen* had previously been divested at Glasgow of all her cargo winches, derricks and everything else that would be of no further use to her during what was to be her future task, which at this time was not known to us. Explosive charges had been laid in two corners of each of her five cargo holds, and electric cables had been run up to her bridge, each one having been connected to the respective charges.

We set sail in convoy together with many other Merchant, Royal Navy and Fleet Auxiliary ships towards the coast of Normandy, having very little idea of what was in store for us. We arrived off Arromanches on the morning of D-Day, and dropped our anchor in line astern of some dozen or so other blockships (as had been designated in our sealed orders). Many others followed us to their respective charted positions.

After a short while we were boarded by the 'wrecking officer' and his party, and without further ado the aforementioned cables running to the bridge were connected to a plunger, the trigger was pressed and Number 307 sank gently on to the bed of the English Channel in about 2½ fathoms of water, close inshore at Arromanches.

Everett and some 30 other crew members remained on board No 307 for a further two weeks, occupying the upper two decks, which were clear of the water. During this time they were on continuous action stations and were attacked several times by German aircraft.

the hulk of the Dutch 'Java'-class light cruiser HMNLS *Sumatra*.

Each vessel was ballasted to draw about 19ft of water, and explosive charges (each containing 9lb of Amatol – between eight and ten per ship, depending on its size) were attached on either side of the hold. These were then connected to an electric firing key on the bridge. One by one the ships were detached to their sinking positions on the instructions of a 'Planter' officer whose duty was to sink each ship in a position decided by him in conjunction with the approved layout plan. At the appointed time and place, the charges were fired and holes were blown 3ft below the water line. The vessel would take on water and settle slowly on to the shallow bottom.

The ships needed to be well locked in, which meant they should overlap with no gaps in between them, otherwise tidal scour would wash away the seabed beneath their hulls. The result would be the undermining of the ships, which would then become unsupported and, as was the case at the American Mulberry A, they would break their backs.

Corncob blockships

The ships used for each of the five beaches included warships, cargo vessels and Liberty ships in their number:

Utah Beach (Gooseberry 1): *Benjamin Contee, David O. Saylor, George S. Wasson, Matt W. Ransom, West Cheswald, West Honaker, West Nohno, Willis A. Slater, Victory Sword* and *Vitruvius.*

Omaha Beach – Mulberry A (Gooseberry 2): *Artemas Ward, Audacious, Baialoide, HMS Centurion, Courageous, Flight Command, Galveston, George W. Childs, James Iredell, James W. Marshall, Olambala, Potter, West Grama* and *Wilscox.*

Gold Beach – Mulberry B (Gooseberry 3): *Aghios Spyridon, Alynbank, Elswick Park, Flowergate, Giorgios P., Ingman, Innerton, Lynghaug, Modlin, Njegos, Parkhaven, Parklaan, Saltersgate, Sirehei, Vinlake* and *Winha.*

Juno Beach (Gooseberry 4): *Belgique, Bendoran, Empire Bunting, Empire Flamingo, Empire Moorhen, Empire Waterhen, Formigny, Manchester Spinner, Mariposa, Panos* and *Vera Radcliffe.*

Sword Beach (Gooseberry 5): *Becheville, Courbet, Dover Hill, HMS Durban, Empire Defiance, Empire Tamar, Empire Tana, Forbin* and *HNLMS Sumatra.*

Bombardons

Bombardons were cruciform floating steel breakwaters, moored end-to-end in a one-mile line and anchored outside the main breakwaters that consisted of the Corncobs (blockships) and Phoenixes (concrete caissons). In this outer deep-water anchorage it was intended that shelter would be found for the unloading of Liberty ships.

Initial tests involved 200ft-long flexible-sided breakwaters with a 12ft beam and a 16½ft draught. The hull consisted of four rubberised canvas envelopes placed one inside the other and enclosing three air compartments, each running the full length of the hull. The envelopes were attached to and supported a 700-ton reinforced concrete keel. Air pressure in the three compartments was adjusted to coincide approximately with the mean hydrostatic pressure on the outside of the respective envelopes, enabling the hull side to move in or out under any temporary imbalance between the two pressures. The flexible-sided breakwater was not adopted for Mulberry owing to the vulnerability of its fabric sides, although it

was important in establishing the validity of the general theory of floating breakwaters.

In June 1943, the first models of the Bombardon floating breakwater were tested, and within two months enough data had been collated to establish the correctness of the theories relating to the rigid-sided type. More than 300 model tests of the rigid type proved that it was possible to construct a floating breakwater which would suppress waves of the maximum size expected during Operation Overlord (Force 6, which equates to approximately 8ft high and 100ft long), using between 1¼ and 2½ tons of steel per foot of breakwater frontage, which represented one-tenth of the cost for any other possible method.

The decision to proceed with a full-scale floating breakwater was taken in Washington on 4 September 1943. As already stated, the Admiralty was to be responsible for their design and construction at several locations including the King George V Dock at Southampton. Over a mile of floating breakwater was soon designed, assembled and then successfully tested in Weymouth Bay in April 1944.

RIGHT Bombardons under construction in a dry dock at Southampton.
(IWM A25813)

Conceived by Lieutenant Commander R.A. Lochner RNVR, each Bombardon unit contained about 250 tons of steel, measured 200ft long, 25ft 1in in the beam, with a 25ft 1¾in hull depth and a 19ft draught. The cross-section was roughly in the form of a Maltese cross. The top half of the vertical arm of the cross was essentially a series of watertight buoyancy compartments made from welded ¼in mild steel plate, while the bottom half and the two side arms were built out of mild steel angles and plate in bolted sections. The bottom and side arms filled with water when the unit was launched and, with a dead weight of 1,500 tons, provided the Bombardon with the necessary mass.

To construct a floating breakwater it was necessary to moor several units in line ahead, much like a series of ships. However, when securing ships to head and stern moorings it was usual to leave a gap between adjacent ships equating to the length of the ships themselves. With a floating breakwater such a practice would result in half the wave energy passing through the gaps between the units and then rebuilding inside the harbour to a wave three-quarters its original height. It was therefore necessary to work with much smaller gaps between units, and after a number of trials the optimum gap was set at 50ft. The gap was successfully maintained by mooring Bombardons in pairs between mooring buoys, with the Bombardons coupled together with twin 18in cable-laid Manila rope.

To further reduce the effects of wave energy inside the harbour it was decided to use two parallel lines of Bombardons spaced 800ft apart, which in practice gave a reduction of wave height to about 30% and a reduction of wave energy to ¹⁄₁₀ of the original incident wave.

In order to moor the large number of Bombardons in close proximity, a system of laying was adopted that ensured accurate spacing on the mooring buoys:

- First, 5in flexible ground wires with 1,000lb sinkers were laid at equally spaced intervals to which wire risers and spherical floats were attached.
- Next, the floats were replaced by mooring buoys, and the seaward and leeward anchor cables were attached.
- Then the seaward leg was secured to two 3-ton and one 5-ton mushroom anchors and one 8-ton concrete clump, and the leeward leg to one 3-ton mushroom anchor.

As soon as the moorings were in position, the Bombardons were attached to the buoys in pairs by their Manila rope connectors. Thanks to their simple layout, 26 moorings were laid and more than two miles of floating breakwaters were assembled off the French coast at Arromanches and Saint-Laurent in six days following D-Day.

The first sections of the floating breakwater sailed with the invasion fleet on D-Day, when they were towed in pairs with the same Manila couplings used to secure them to the mooring buoys acting as towing links between the two Bombardon units. The breakwaters were moored in 11 to 13 fathoms of water, which gave sufficient depth inshore for Liberty ships to anchor.

ABOVE A section of Bombardons in operation as the outer breakwater to the Mulberry harbours. *(USNA)*

Phoenix

Phoenixes were enormous rectangular
cellular reinforced-concrete caissons that
were used in conjunction with Corncobs and
Bombardons as part of the outer breakwater
when partially sunk in 5½ fathoms of water
(1 fathom = 6ft depth). There were six sizes,
which varied from 1,672 tons (D1 type) to 6,044
tons (A1 type) each, depending on the depth
of water in which they were to be planted. The
Phoenix units were fabricated in England and
were given swim (upswept) ends to make them
towable, each unit requiring two large (759hp)

tugs to tow it across the Channel at an average
speed of three knots. Four Phoenixes, with
most of their crews, were lost in the assault
phase. Two were lost to enemy action (E-boat
torpedoes and a mine) and two were lost in
bad weather.

Design

Designed by War Office drawing office
staff under the supervision of Captain W.J.
Hodge, and with the assistance of consulting
designers from outside, the Phoenix design
was kept deliberately simple. The A1-type
Phoenix, as originally designed, had 10

separate watertight compartments, five along each side of a longitudinal centreline division. Each compartment had two valves for flooding (they allowed the Phoenix to be sunk quickly and to control its stability while sinking) – a 12in and a 5in sited 2ft and 12ft above the bottom respectively. Three dwarf walls to break up the free surface inside the hold helped the caisson to sink on an even keel. Reinforcing was almost entirely with straight iron bars, and large openings were left in the internal cross-walls, partly to reduce weight and to keep the centre of gravity low. To facilitate towing, upswept swim ends were incorporated in the design.

ABOVE For air defence, a 40mm Bofors gun was mounted on top of some Phoenix caissons. *(USNA)*

LEFT On Hayling Island in Hampshire the area around the Ferry Boat Inn was used to construct Phoenix caissons. One of these caissons developed a fatal crack and was eventually abandoned on a sandbank in Langstone harbour. Because it was an obstacle to shipping, the caisson was refloated and towed to safety on Sinah sands where it was again sunk. *(Author)*

BELOW Phoenix B2 schematic. *(Copyright unknown)*

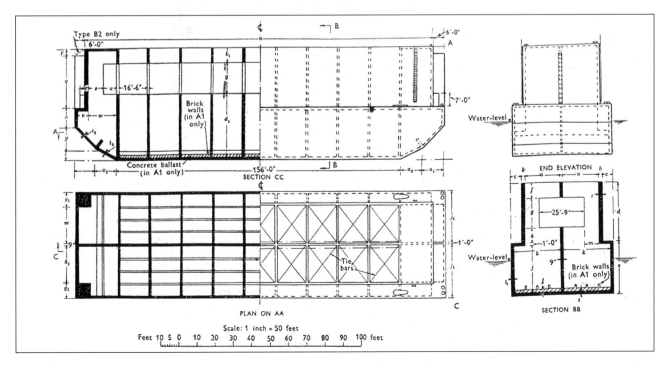

Crew spaces were provided at the ends, where the end bays, or compartments, were roofed over and wooden floors provided at a suitable depth below the roof. However, the crew spaces were only available when the caissons were afloat, because when sunk the floors were awash.

A gun platform with a 40mm Bofors anti-aircraft gun and a rough shelter for the gun crews was provided only on the tops of the A- and B-type caissons.

The cellular construction was open to the sky, and narrow walkways across the cells gave crew access to the fore and stern ends. Six-foot insets (or gangways) ran round the lower edges of the sides and end walls of the larger caissons, originally for use during construction, but also giving accessible space on completion for the installation of towing bollards and fairleads at each end, and stowage space for the heavy cylindrical concrete clump anchors. They also made convenient boarding platforms.

Later, these insets (or gangways) were provided with four suction pump holes (fitted with wooden covers) on each side gangway.

They accommodated the pipes for 6in suction pumps that stood on the gangways for pumping out the caissons after temporary sinking.

Construction

Of the original 147 Phoenix caissons required by D-Day all except the 24 B2 units were laid down in the sites mentioned below, from where they could be floated out when either fully or partially completed. In cases where partially completed craft were floated off, construction berths were used twice in succession. It would have made sense to build them all in dry docks but the Admiralty had other ideas, commandeering the dry docks for shipping repairs and the construction of their own Bombardons, so the War Office had to find alternative sites.

Graving docks were made available at various locations around England, while a number of caissons were constructed on beaches near Portsmouth and launched down slipways, or built in temporary excavated basins along the Thames estuary, from where they were floated out when partially complete, and the work finished afloat.

BELOW A view along Stanswood Bay, near Lepe in Hampshire, showing some of the B2-type concrete caissons for the Mulberry harbours constructed at the site between 1943 and 1944. The caisson second from the camera has collapsed in its frame during launch. *(IWM H35554)*

Raising the Phoenix
Freddie Downing, Bovis Ltd

At the age of 30 Freddie Downing was involved in building Phoenix caissons on the south coast for construction firm Bovis in 1943–44. He describes how dangerous it could be working on these highly unstable floating platforms. After the war Downing became assistant managing director of Bovis Construction.

When I started off in Portsmouth the first of our structures was going up in dry dock. They were being built to 25ft in the dry dock, then flooded and floated out to a wet dock. That's where I took possession of these boats, as we called them.

I then put on another 30ft while they were floating and unstable. Bovis pumped the concrete in the dry dock. On the water we used hundreds of barrow runners. Because the caisson was floating we had to be able to take the concrete to where it was needed to keep the thing stable, and the barrows were better for that. The system was not foolproof though.

I was moved on to some units being constructed in Southampton where Bovis was helping Pauling out. We took over with 15ft to go. When we got there, the units were tied up alongside a jetty on Southampton Water, exposed to the sea. Southampton Water has a 17ft tide twice a day, so we were up and down all the time.

Bovis's best foreman, a chap called Page, decided I think to show how quickly he could get concrete laid on his side of the boat. He did it too well, because suddenly there was a lot of shouting and the caisson started to

list to one side, far enough to let water in through the scaffolding holes. A plumber, Dudley Wheeler, shot down inside the unit to open up some penstock valves on the other side in an attempt to right it (he won the BEM for his bravery). This straightened the unit up, but too late and water was still pouring in. We stood on the dock side and saw it sink to the bottom.

The bottom of the dock was sharply shelving and Downing watched as the 40ft-high concrete caisson launched itself out into Southampton Water. When it finally came to rest, the water level reached the top of the shuttering and all that could be seen was bits of scaffolding tube sticking up out of the sea.

ABOVE Concrete caissons under construction in dry dock. *(IWM A25791)*

LEFT A Phoenix caisson, flooded and floated out to a wet dock for completion. *(Author's collection)*

Phoenix construction was carried out at the following sites:

- Middlesbrough – graving dock.
- Goole – graving dock.
- Southampton – graving dock.
- Portsmouth – entrance lock and floating dock.
- Tilbury (Port of London Authority – PLA) – graving dock.
- East India (Import) (PLA) – dried-out wet dock.
- South Dock (Surrey Commercial) (PLA) – dried-out wet dock.
- 12 basins were specially excavated and prepared along the banks of the River Thames where 40 caissons could be laid down at one time (98 caissons in all were laid down there).
- 24 B2 units were constructed on slipways along the south coast for broadside

Materials used to build Phoenix caissons

Original programme:
Concrete – 410,000cu yd formed from the following materials:
    Sand – 179,000cu yd
    Aggregate – 358,000cu yd
    Cement – 129,000 tons
Mild-steel-bar reinforcement – 29,570 tons

Supplementary caissons:
Concrete – 132,000cu yd formed from the following materials:
    Sand – 57,500cu yd
    Aggregate – 115,000cu yd
    Cement – 41,400 tons
Mild-steel-bar reinforcement – 19,600 tons

Principal dimensions of Phoenix caissons

Unit	Height	Length	Breadth at WL	Displacement	Draught
A1	60ft	204ft 0in	56ft 3in	6,044 tons	20ft 3in
A2	50ft	204ft 0in	56ft 3in	4,773 tons	16ft 4in
B1	40ft	203ft 6in	44ft 0in	3,275 tons	14ft 0in
B2	35ft	203ft 6in	44ft 0in	2,861 tons	12ft 5in
C1	30ft	203ft 6in	32ft 0in	2,429 tons	14ft 3in
D1	25ft	174ft 3in	27ft 9in	1,672 tons	13ft 0in

launching at Stokes Bay, Gosport; Lepe Beach and Stone Point on the Solent; and at Langstone harbour near Hayling Island.

The whole Phoenix construction programme was undertaken by 24 large contracting firms, of which many, like Wimpey and Costain, became household names in the post-war UK construction boom. Contractors chose their own methods of construction, which depended very much on the facilities available on site, as well as plant, shuttering and other materials, and of course the supply and quality of labour.

A 20,000-strong workforce toiled day and night from December 1943 (this figure does not include those who were employed in the fabrication of parts and manufactured materials produced off the sites). In part because of unskilled labour and concrete production processes that required maximum speed, the quality of the caissons and the pace of construction varied greatly. Another problem facing constructors was that experimentation was still going on with the basic design, which meant that any modifications arising from the experiments had to be incorporated during construction – there was a 50% increase in the estimated use of steel, and extra work not included in the original programme involved the addition of gun towers and more deck gear. Even so, the original programme was completed in 150 days, with the first caisson arriving at Arromanches on D-Day+1.

It was originally intended that all Phoenix units were to be kept afloat until required, but sufficient moorings were not available, so suitable sites had to be found to sink them temporarily, and then when the time came to raise them they were pumped out. A number of sites were chosen off the south coast between Selsey Bill and Bognor Regis, and at Littlestone near Dungeness. The stretch between Selsey and Bognor was good, and all the 80 caissons parked there were safely raised without mishap (although one A1 unit was irretrievably damaged when it was sunk for a second time). The Dungeness site proved bad on account of tidal scour, and six small caissons were lost to use there.

The parking of caissons had an unexpected spin-off in that it afforded the crews valuable pre-D-Day experience in handling and sinking.

Riding the Phoenix
Mick Crossley, 416 Battery, 127 LAA Regiment, Royal Artillery

ABOVE Mick Crossley, Royal Artillery. *(Mick Crossley)*

Mick Crossley was one of dozens of Royal Artillery gunners who were detailed to man Bofors anti-aircraft guns mounted on top of the Phoenix caissons. He quickly discovered that life at sea on top of a floating concrete box was anything but comfortable.

We were taken out from Folkestone harbour in commandeered fishing boats, a detachment to each boat, and introduced to the 6,000-ton concrete Phoenix caissons. On the way [across the Channel] we noted that the caissons, each over 200ft long and 30ft high from the water level, were dotted along the coastline, laying about a ¼-mile offshore, and they had Bofors guns mounted upon a central tower. We climbed from the fishing boat on to a lower ledge of the caisson and then had to climb 25–30ft up a vertical iron-runged ladder to reach the top and deck. I remember there were one or two members who were too scared to make the climb, which was understandable, and they had to be roped up.

Once on top we found the caisson was hollow, divided into about 16 compartments, each filled with water to the level of the tide at the time. We found a small deck at each end with a catwalk leading down the middle of the caisson to the gun tower. We were to live on these monsters for a week at a time to get used to them. The designers had made a small concrete room at one end with a small window for any gun crew to shelter in. We found that sleeping was impossible there and we all had claustrophobia, preferring to bunk down underneath the gun tower.

Eventually our big day arrived. First, two Royal Engineer sappers and two Royal Navy ratings came aboard to accompany us across the Channel – the Engineers to man the generator and 'pump out', and the naval men to assist in the positioning of our caisson at Arromanches. Also to join us were three men from our Battery HQ, just for the crossing. We were off, but it was to be a long and very slow journey.

RIGHT Phoenix B90 at sea and under tow by three ocean-going tugs. The letter 'B' denotes the caisson is destined for the British Mulberry B at Arromanches; the number '90' is its position in the breakwater – in this case at the western end of the harbour. The letter 'M' on the funnel of the tug on the left shows it is part of the Mulberry tug fleet. *(USNA)*

Spud piers

With sheltered water having been provided for, the next most important structure was the means for the rapid unloading of stores and vehicles. The pier heads (known as Spud piers because of their four long legs, or spuds) were used for unloading coasters, LST and LCT. They consisted of a rectangular steel floating platform 200ft long, 60ft in the beam, and about 11ft deep with a leg, or spud, at each corner standing 90ft tall, which could be lowered to the seabed when the pier head was in position.

The deck had non-skid treads welded to its surface. Seatings for bearings for the floating bridge spans (Whales) were provided at eight deck locations – three on each side and one at each end. Two longitudinal and seven transverse plate girder bulkheads divided the pontoon into 24 watertight compartments. Deck equipment was much the same as that found on any normal ship and included winches, capstans, derricks, anchors and bollards.

The spud framings near the ends of the pontoons, which included the spud guides, were of heavy welded construction. Each of the four spuds was made from steel plate, measured 4ft square and 89ft long, and the spud feet were 8ft square.

Crew accommodation was for one officer, six NCOs, and 15 men, with cooking, heating and toilet facilities.

Electric power for the pontoon was generated by two 57kW diesel engines and generators. This power was mainly used for operating the spuds, but it was also used for deck equipment. In the event of breakdown, power could be fed to, or drawn from, adjacent pontoons.

The spud legs were remotely operated and independently controlled by an operator from a control room in the bridge at one end of the pontoon. He raised or lowered the hull periodically to suit the tide as it was rising or falling. Each leg was independent, which meant that whatever unevenness was found on the seabed below, the legs would find their own level and firmly anchor the horizontal pier weighing about 1,000 tons (1,760 tons with its attached intermediate pontoon).

Each spud was moved by two twin wire-ropes that were attached at one end to a barrel driven by a 20hp winch with a reduction gear, and at the other to the spud keeper anchorages, to raise or lower the spuds at about 2½ft/min (which was well in excess of tidal requirements). An arrangement called the spud down-haul rope lifted the hull above the free flotation level by pressing the spud foot down on to the seabed (although the spuds were not designed to lift or hold the pontoon completely out of the water). In very calm weather the four spuds allowed the hull to float freely up and down with the tide, but normal operation was for the hull to be lifted 6in above the free flotation level.

LST pier heads

To assist the LST and LCT in discharging their cargoes, two Spud piers were placed at right angles to one another in the shape of a letter 'T' and were equipped with a buffer pontoon on each side of the shaft of the 'T', providing an artificial beach that the disembarking vehicles could use to drive up on to the deck of the pier. To allow LST to unload vehicles from off their top-decks as well as through their bow doors, a high-level hinged ramp was attached to the platform on the Spud pontoon deck. The ramp was on a slope of 1 in 3½ and could be raised, lowered or moved longitudinally by means of chain blocks suspended from gantries. This meant that by means of the buffer pontoon and high-level ramp an LST could be unloaded in 17 or 18 minutes. Two LST could berth and discharge their cargoes simultaneously from the T-shaped LST pier head.

The construction of the pier heads was carried out at Leith, the military port of Cairnryan, and at Conway in North Wales by Joseph Parks & Son of Northwich, Cheshire. The launching of the completed craft was undertaken by Hollway Bros (London) Ltd. After fitting out, the pontoons were towed to Southampton for additional work to be carried out to complete their mechanical

and electrical installations. Once finished they were tested and then towed away to the assembly area at Selsey.

Once in situ on the Normandy coast, Spud piers were strung together 160ft apart in line to form a long pier head for unloading stores from cargo vessels. To increase the length of the berthage, in between each pair of Spud pontoons there was a concrete intermediate Pier Head Pontoon (PHP) and a telescopic Whale bridge span.

ABOVE An LST comes carefully alongside the pier head at a speed of about three knots. *(USNA)*

LEFT A high-level hinged ramp was attached to the platform on the Spud pier deck, enabling LST deck cargo to be unloaded while the main tank deck was being emptied. *(USNA)*

Concrete intermediate Pier Head Pontoons (PHP)

PHPs were rectangular compartmented pontoons with flared sides, and vertical ends, but with the bottoms faired at the ends for ease of towing. Each unit measured 80ft long by 56½ft wide, with a draught of 6ft 9in unloaded. It displaced 710 tons light, 1,660 tons loaded, and when fully loaded its freeboard could be as little as 1ft, although this was not a usual working condition. The PHP was attached to one Spud pier, with a telescopic bridge span connecting the free end of the PHP and the next Spud pontoon. There were 18 watertight compartments inside a pontoon, three of which were reserved for stores and one for living accommodation if needed.

Exacting design requirements stipulated that a PHP should be able to carry a 40-ton military tank moving anywhere on deck, in addition to 20-ton axle loads from other vehicles travelling in any direction on the deck. It should remain stable and afloat even with two 40-ton tanks on one side of the pontoon deck, while the remainder of the deck was unloaded, and it should be able to withstand seas up to 8ft high and 120ft long.

Because of the shortage of steel, PHPs were constructed from reinforced-concrete beams and panels. The concrete bottom panels and beams, each of which measured 13ft 3in long x 3ft wide x 2½in thick, ran longitudinally from bulkhead to bulkhead and were designed (and reinforced with ⅜in bars) to withstand water pressure, but not grounding on a beach or shallow seabed. The side-wall panels were similar in form and construction to the bottom panels, while the longitudinal and transverse bulkheads were built of similar units to the walls, the only difference being one of arrangement.

Construction of the PHP deck was unlike the rest of the structure because it had to be able to take heavy rolling loads in contact with its surface. The beams ran longitudinally every 3ft 6in (as in the bottom) and were supported by the cross-bulkheads and the end walls. A new and heavier type of panel was fabricated that combined slab and beams in one precast unit. It took the form of an inverted trough 3ft 6in wide and 17in deep. The deck slab section was 4in thick and reinforced with bars. Each deck panel weighed two tons. The jointing of adjacent panels was sealed with mortar where necessary, but the panels were usually butted together to form a robust deck area.

RIGHT This intermediate Pier Head Pontoon (PHP) is attached to Spud pier head 589. In the foreground a telescopic section of Whale roadway is prepared to be attached to the PHP with the aid of an air-filled erection tank (at lower left). *(IWM B5733)*

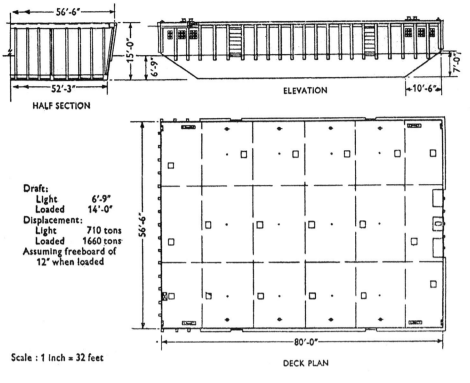

Draft:
Light   6'-9"
Loaded   14'-0"
Displacement:
Light   710 tons
Loaded   1660 tons
Assuming freeboard of
12" when loaded

Scale : 1 inch = 32 feet

HALF SECTION

ELEVATION

DECK PLAN

LEFT General arrangement drawing of a PHP. *(Author's collection)*

To protect the concrete structure from berthing knocks, 8in x 5in Columbian pine vertical fenders were attached at 3ft 6in intervals to the sides and ends, and there were 9in x 9in horizontal fenders at deck level. Four substantial timber access ladders were added to the fendering, two per side of the PHP.

Deck fittings comprised 8in cast steel bollards (one at each corner of the deck) with cast steel fairleads; deck rings for general use (four along each side); cast-iron watertight 18in square manhole covers (one per compartment); steel bearing plates for ends of bridge girders; pivot plate for anchoring bridge to pontoon; steel clench plates to take wire rope attachments to Spud pontoons; steel rubbing strips at each end of the deck to take the ends of steel ramps to Spud pontoon and telescopic bridge sections.

Each PHP was attached to its adjacent Spud pier head by wire strops and springing. The other end of each PHP was restrained laterally by wire ropes passed diagonally underneath the telescopic bridge span and connected to the next Spud pier.

As mentioned above, because of the shortage of steel, PHPs were initially made of metal reinforced-concrete panels, but because of their susceptibility to damage from berthing

shocks and inadvertent grounding they were later replaced by steel pontoons. At Leith 16 intermediate steel pontoons were made.

BELOW That this 850-ton PHP has withstood 70 years of battering by the English Channel is testimony to the strength of the original design. *(iStock/Brett Charlton)*

Intermediate Pier Head Pontoon construction yards and contractors

Site	No constructed	Contractor
Marchwood	6	Wates Ltd
Beaulieu	6	Wates Ltd
Barking	4	A. Monk & Co. Ltd
Rainham	2	The Trussed Concrete Steel Co. Ltd

Buffer pontoons

Designed by William Wilson and Frederick Sully for War Office department Tn5, the buffer pontoon was a floating ramp of welded steel plate construction with a cellular interior, designed to absorb the impact from the bows of an LST when berthing. It also enabled vehicles to drive off the LST and up on to the Spud pier under their own power.

Buffer pontoons were built by Holloway Bros (London) Ltd at Conway and at two open sites on the south-east coast. These odd-shaped structures were enclosed by a rectangle 76ft 4in by 65ft, had a launching weight of 469 tons, and were side-launched from their building berths.

Owing to the acute rake of the forepeak of the LST, the bottom of the bow doors was prone to catching on the ramp of the buffer pontoon after the vessel had come to rest. The doors could only be opened if the vessel was pulled back from the ramp. One solution to the problem was to lower the nose end of the pontoon to allow clearance for opening the bow doors. This was made possible by two electrically operated low-pressure compressors for flooding (and afterwards de-watering) internal ballast tanks to adjust the trim of the pontoon by 3ft 6in. With the vessel's bows also depressing the pontoon the bow doors could open. The buffer pontoons were used at the British Mulberry B only, the Americans deciding to burn off the corners of the bow doors of their LST so they could open them on the ramp.

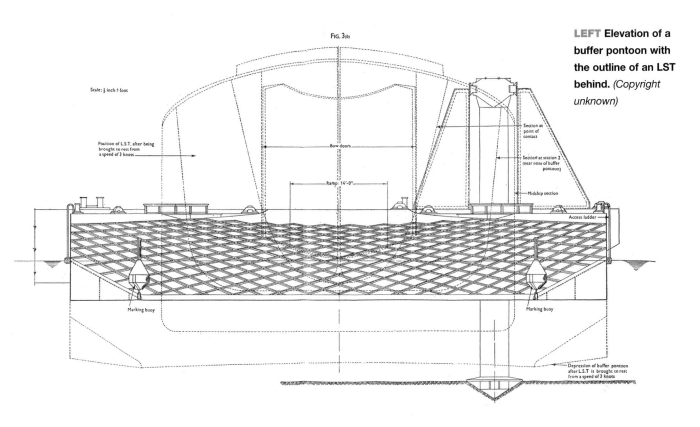

Fig. 3(b)

Scale: ⅛ inch-1 foot

Position of L.S.T. after being brought to rest from a speed of 3 knots

Bow doors

Ramp: 14'-0"

Section at point of contact

Section at station 2 (near nose of buffer pontoon)

Midship section

Access ladder

Marking buoy

Marking buoy

Depression of buffer pontoon after L.S.T is brought to rest from a speed of 3 knots

LEFT Elevation of a buffer pontoon with the outline of an LST behind. *(Copyright unknown)*

BELOW A section through the centreline of a buffer pontoon. *(Copyright unknown)*

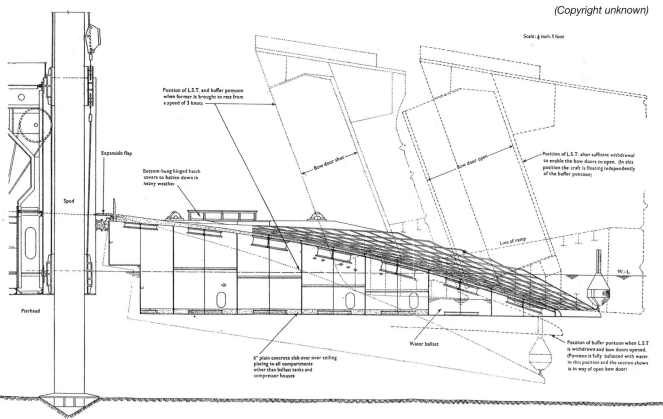

Scale: ¼ inch-1 foot

Position of L.S.T. and buffer pontoon when former is brought to rest from a speed of 3 knots

Expansion flap

Bottom-hung hinged hatch covers to batten down in heavy weather

Bow door shut

Bow door open

Position of L.S.T. after sufficient withdrawal to enable the bow doors to open. (In this position the craft is floating independently of the buffer pontoon)

Spud

Line of ramp

W.-L.

Pierhead

Water ballast

Position of buffer pontoon when L.S.T is withdrawn and bow doors opened. (Pontoon is fully ballasted with water in this position and the section shown is in way of open bow door)

6" plain concrete slab over over ceiling plating to all compartments other than ballast tanks and compressor houses

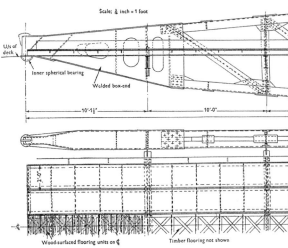

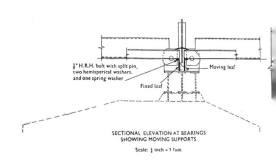

Whale flexible roadway carries vehicles ashore on to Omaha Beach. Note the sign reminding users of the 25-ton weight limit. *(USNA)*

Whales

Whales were floating roadways that connected the Spud pier heads to the land. They enabled coasters, LST and LCT to be discharged direct to the shore. Built to ride out the rough weather, these roadways were made from 80ft-long flexible steel bridging units, which were mounted on pontoons made either of steel or concrete, known as Beetles. Many of these Whale bridge spans from Mulberry B at Arromanches were later used to repair bridges that had been destroyed during the war in Belgium and Holland as well as in France.

The challenge to the bridge designer, William T. Everall, was to design a pontoon bridge able to ride a rough sea without overstressing any of its components. Solutions had to be found to accommodate the rolling, pitching and torsional (twisting around its axis) motions of the structure caused by wave action, because if not controlled they would tear the bridge apart. Calculations and tests established a requirement to allow each individual bridge span an angular movement of 24° relative to another, and a torsional displacement of 40° along the length of each span.

His solution was ingenious:

- The main girders were lozenge-shaped to enable the chord material to withstand both bending stresses and longitudinal forces.
- The centre sections of the girders were of built-up welded and bolted lattice construction.

- The ends were of a welded box construction incorporating a spherical bearing.
- Longitudinal forces exerted by the tendency of the structure, under the influence of wave action, to rotate about a vertical axis and move bodily athwartships were transmitted from one bridge span to another through thrust bars built into the bearings.
- Tensile loads were resisted by means of steel cable grommet strops secured to the bearing housing after passing through a very large sheave worked into the girder end.

Also involved in the early development work of the bridge spans was Everall's principal assistant, Allan Beckett, who designed the Kite anchor (see pages 94–95).

The standard flexible bridge span for Mulberry was designed for Class 25 loading, although the scheme included a stronger version to carry Class 40 loads. The 80ft-long span was of welded and black-bolted construction, with a 10ft-wide steel deck with a non-slip tread and wheel guards supported on cross girders.

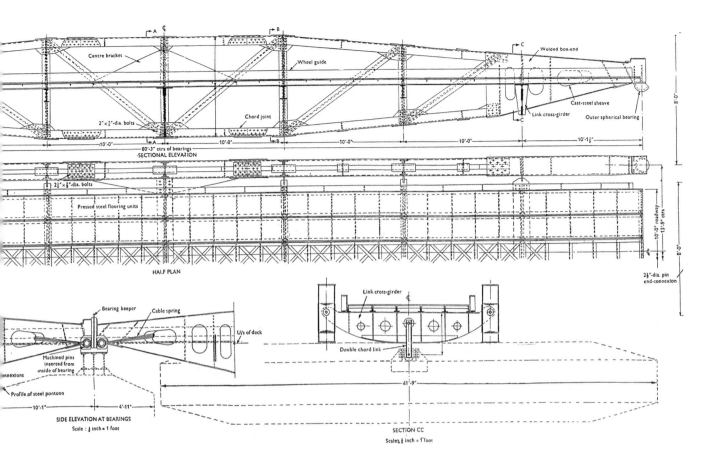

Centre bracket

Wheel guide

Welded box-end

2" × ⅞"-dia. bolts

Chord joint

Cast-steel sheave

Link cross-girder

Outer spherical bearing

SECTIONAL ELEVATION

10'-0"

10'-0"

10'-0"

10'-0"

10'-1½"

80'-3" ctrs of bearings

8'-0"

2½" × ⅞"-dia. bolts

Pressed steel flooring units

HALF PLAN

10'-0" roadway

13'-9" ctrs

8'-0"

2½"-dia. pin end-connexion

Bearing keeper

Cable spring

U/s of deck

Link cross-girder

Machined pins inserted from inside of bearing

Double chord link

onnexions

Profile of steel pontoon

10'-1"

4'-11"

41'-9"

SIDE ELEVATION AT BEARINGS

Scale : ¼ inch = 1 foot

SECTION CC

Scale ½ inch = 1 foot

ABOVE A standard span for Deep-Water Floating Bridge Mk I Whale. *(Copyright unknown)*

A telescopic bridge span was also designed which, like the standard span, was flexible, but was self-adjustable in length between 71ft and 80ft. It was used at intervals along the length of the roadway to overcome four problems arising from the rigidity of the structure. It enabled the bridge to:

- Adjust itself in length to compensate for the height of tides.
- Lengthen or contract to allow for variations in the angle of the roadway caused by low or high tides.
- Compensate for lateral movements under rough sea conditions that caused the bridge to expand and contract, thereby minimising the heavy stresses on the flexible couplings joining the spans.
- Link pier heads together without the need for accurate placement of each pier. The telescopic spans could be adjusted to the precise distance between adjacent piers.

ABOVE A telescopic bridge span attached to an LST Spud pier head, with an LST alongside. Wheel guards and non-slip treads on the roadway can be clearly seen in this photograph. *(USNA)*

Beetles (PP)

Beetles were bridge-carrying pier pontoons (PP) that supported the Whale piers or roadway sections. They were made from concrete or steel and were designed to function in shallow water or fully aground (the steel type being used over rock). Pontoons were to be capable of being towed 100 miles across the Channel carrying a bridge span each, or in pairs two-abreast carrying two bridge spans between them placed one above the other.

Concrete pontoons

Steel was considered the most suitable material, but owing to a severe shortage of steel plate and steel workers a substitute became necessary. As a programme of reinforced-concrete barge construction had been successfully in progress since 1942, the possibility of adapting the pre-cast principle was looked at. Consideration was given to reducing the weight of the concrete to make the pontoon easy to tow; making the skin waterproof; and incorporating fendering to protect the pontoon against being holed. A series of experiments and trials were carried out before the best design was arrived at that fulfilled the requirements for towing at sea and carrying the bridge spans once in situ.

PP Types 5, 6 and 7 were built of precast vibrated concrete panels that were assembled rather like you would an Airfix plastic model kit. The Beetles were symmetrical in shape but asymmetrical in their cross-section, the reason for the latter being to minimise the chances of bumping between the pontoon's edge and the bridge girders it was supporting when the sea was choppy.

Internally the pontoon was subdivided into six watertight compartments by five transverse bulkheads formed of 2in-thick precast concrete panels. Some 65 precast 1¼in-thick panels

RIGHT A general-arrangement drawing of a reinforced-concrete pier pontoon Type PP6. *(Copyright unknown)*

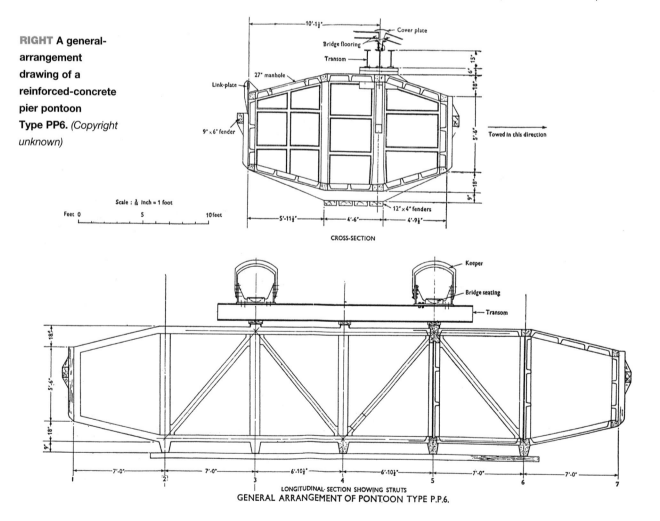

GENERAL ARRANGEMENT OF PONTOON TYPE P.P.6.

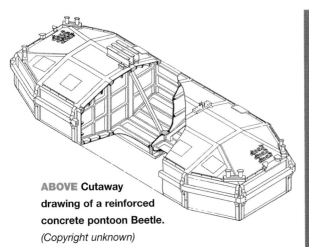

ABOVE Cutaway drawing of a reinforced concrete pontoon Beetle. *(Copyright unknown)*

PP5 Type	
Designed to carry:	Bridge and equipment (35½ tons) plus 25-ton military tank, plus reserve buoyancy of 23 tons
Dimensions:	41ft 9in x 14ft 11½in x 8ft deep
Weight:	43½ tons
No built:	31
PP6 Type	
Designed to carry:	Bridge and equipment (35½ tons) plus 25-ton military tank, plus reserve buoyancy of 23 tons
Dimensions:	41ft 9in x 15ft 3in x 9ft 7in deep
Weight:	46 tons
No built:	327
PP7 Type	
Designed to carry:	Bridge and equipment plus 40-ton military tank
Dimensions:	41ft 9in x 18ft 9in x 9ft 10in deep
Weight:	60 tons
No built:	126

made up the 26 facets and five bulkheads per pontoon, the number of precast units comprising 36 different types of panel. (The PP Type 7 was made up from 66 panels of 37 types.)

The concrete panels were prefabricated on vibrating tables at the yards of precasting firms. All edges of slabs were treated with a water spray at a few hours old to ensure a proper jointing surface. The precast panels were then cured with water for the first three days and were usually transportable by day four. One-ton mobile cranes were used for erecting the slabs, which were held in position by scaffolding until the in-situ concrete work was complete. It took four to five days to assemble and complete a Beetle before launching. Snag-checking and fitting-out took place after launching, with special gangs trained to find and repair any leaks in the structure.

Steel pontoons

Steel pontoons were reserved as far as possible for the shore end of the piers. With a length of 42ft, a breadth of 15ft and a depth of 8ft, they were formed of six separate all-welded units of ³⁄₁₆in steel plate bolted together through flanges. The bottom plating was ¼in thick. Bulkheads underneath the bridge bearings (which carried the Whale roadway) were heavily stiffened and extended by 9in externally below the bottom of the pontoon to form bumpers, which were shod with hardwood.

Deck fittings included hand-railing, 9in built-in bollards, and combined fairleads and deck stoppers at each end. Between the bridge bearings and the bulkhead projections

above deck was a transom formed of three 15in x 16in RSJs, which carried the flooring connections and formed a springy medium to damp out small movements between pontoon and bridge spans.

For sections of the pontoon bridge to be grounded over a rocky seabed, steel pontoons

BELOW A concrete Beetle is launched into the Solent at Marchwood. Note the two men standing on top of the float. *(IWM A25810)*

with four spud legs were used. A hand-wheel locking device allowed the legs to be modified and secured in any position, while an adjustable 'Camel's Foot' allowed accurate alteration to be made on the seabed. For towing, the legs were raised to their uppermost position.

The first tows of Beetles and Whales for Mulberry B started arriving on 9 June. When under tow across the Channel the sea-keeping qualities of the Beetles were poor and they were prone to flooding. Many were unable to survive the crossing in anything but the calmest of seas. Because the pontoons were towed broadside-on, the concrete deck bollards were placed under excessive strain and the tugs frequently pulled them out of the deck completely. On occasions the erection tanks came adrift under tow when making way at anything more than the regulation speed of 3–4 knots, which caused the front span to sink. The loss rate was high, with some 40% of the tows (pontoon and bridge spans) being lost at sea.

Keeping it all together – The Kite anchor

'One might ask what the difference is between the Kite anchor and any other mud hook. The difference is the same as that between an aeroplane that can taxi but not take off, and one that can fly.' Major Allan Beckett

LEFT Major Allan Harry Beckett, Royal Engineers, inventor of the Kite anchor. *(Tim Beckett)*

The key to the Mulberry harbour's success lay in the highly effective Kite anchor that dug itself into the seabed to hold the floating harbour in position.

The Whale bridges and their pontoons had to be moored securely in the waters off the Normandy beachhead to make them safe for use by wheeled transport in all weathers and tidal conditions. To achieve this, pontoons needed to be held in position within a range of movement of about 6in by an anchor and cable arrangement with a holding power of 30 tons. Existing anchor designs proved unsatisfactory because they dragged along the surface of the seabed (instead of anchoring), and they often broke.

Variable-tension long steel mooring cables attached to an ingenious device called a Kite anchor proved to be the solution to the problem. The cables could be adjusted to compensate for variations in loads of between 5 tons at low water and some 12 tons at high water. Wave action could increase these loads up to about 25 tons, but the kite anchor was able to hold up to 30 tons.

This is how the Kite anchor was described by its inventor, the civil engineer Allan Beckett:

The principle of the whole anchor gliding to a depth increasing with pull… is the fundamental of the Kite anchor. The balance of forces to achieve this end is the same as in a kite, which glides to a greater elevation with increased pull on the line. While other anchors scratch around the surface [of the seabed] with the hope

LEFT An actual-size reconstruction of a Kite anchor. *(Tim Beckett)*

By D+5, four Whale tows (including 600yd of roadway) had been lost.

On 13 June, Rear Admiral William Tennant (Rear Admiral Mulberry and PLUTO), who was placed in charge of transport, assembly and the setting up of Mulberry on the Normandy coast, decided to reduce the tows from six spans to just three and for them to make the sea journey in daylight rather than by night. For the Royal Engineer sappers who were detailed to man the spans at sea it was the short straw, offering

a cold and wet voyage with little more than a tarpaulin for shelter.

Beetles were moored in position using wires attached to Kite anchors, which were also designed by Major Allan Beckett, Royal Engineers. These anchors had such high holding power that very few could be recovered at the end of the war. The Navy was dismissive of Beckett's claims for his anchor's holding ability, so Kite anchors were not used for mooring the Bombardons.

of finding something to catch on to, the Kite anchor glides downwards until it finds the increased resistance that is sure to be there.

To enable a Kite anchor to be laid quickly, it had to be lightweight (when produced it weighed 6cwt) and it had to be easily handled by small craft capable of operating in shallow water.

A small engineless 20ft-long craft, called a mooring shuttle, was used to drop the anchors. Made of two plywood floats braced together about 3ft apart, carrying a drum in the middle containing 1,200ft of spooled mooring wire, the shuttle carried an anchor at each end. The shuttle was towed to its dropping position by a shallow-draught launch.

Some 2,000 Kite anchors were manufactured for use in the Mulberry harbours.

ABOVE Mooring shuttles with Kite anchor cables, stowed on a section of Whale floating roadway, ready to be deployed for mooring the Beetle pontoons. *(USNA)*

LEFT A Kite anchor general-arrangement drawing. *(Copyright unknown)*

Deployment

On the eve of D-Day the marshalling of the Mulberry components for the voyage across the Channel was almost complete. Most of the Phoenix units and Spud pontoons and a small number of Whale units were at Selsey; most of the Whale roadway tows were at Peel Bank on the Solent; the reserve Phoenix units and a few Whales were parked at Dungeness and a handful of spare Phoenix units were moored in the Thames estuary. The three Corncob convoys had sailed from Scotland, with the first two arriving at Poole on 5 June

and the third on D-Day itself. The Bombardons were at Portland, where they had been since 15 May.

Beginning on the afternoon of D-Day, and for the next 48 hours, the various Mulberry elements set sail from the south coast for the shores of Normandy, where Royal Engineers and American construction battalions began the gruelling task of assembling the harbour units.

Mulberry A

The first Corncob was planted on 7 June, construction of the Bombardons began on 6 June and the first Phoenix was sunk on 9 June, which saw the whole Gooseberry finished on 11 June. Two gaps were left in the breakwater as additional entrances for landing craft, but as time would tell this considerably reduced the effectiveness of the breakwater. By 18 June, two piers and four Spud pier heads had been completed and were working.

Mulberry B

All Corncobs had been sunk in position by 10 June and all Bombardons were in position by the 13th. By 18 June, 25 Phoenix caissons had been planted. As for the Whale piers, the east pier had been completed by 14 June, with four Spud pier heads added by the 18th, as well as parts of the centre pier and the LST pier.

The Great Storm

Both harbours were almost fully operational when disaster struck on 19 June, as what was believed to be the greatest storm to hit the English Channel and the Normandy coast in two generations arrived. On that day a stiff northerly wind blew up, with a choppy sea and the wind veering from north-north-west to north-north-east, ranging from Force 6 to Force 8. Such a storm had not been recorded in the Channel during the summer for over 40 years. The Mulberry harbour equipment had been designed to withstand the typical weather conditions in the Channel during the summer months, but it was now subjected to a brutal test for which it had not been designed, and at a stage before they were complete.

The storm did immense and irreparable damage to Mulberry A, where 21 out of 28 Phoenix caissons were destroyed and sunk, with four more badly damaged, and where Bombardons broke adrift and Whales and piers were smashed. The backs of seven Corncobs were broken, and serious scour of the seabed resulted in some of the blockships sinking by a further 6ft to 8ft.

Mulberry B suffered much less damage than its American counterpart, but the Bombardon

LEFT A Whale roadway lies twisted alongside storm-damaged Spud pier heads. *(USNA)*

breakwater broke up completely, and most of it came ashore to the west of the harbour. Six Phoenix units were destroyed and there had been slight movement of some of the Corncobs. Three pier heads were damaged, while craft that had dragged their moorings were driven on to the Whale piers, causing considerable havoc.

LEFT The redundant warship HMS *Centurion*, one of the Corncobs, is battered by heavy seas in the Great Storm that raged in the Channel between 19 and 21 June. *(USNA)*

Tombola keeps the tanks full

To provide enough fuel, oil and lubricants for Allied vehicles and aircraft in Normandy in the two weeks following D-Day, the Eastern (British) Task Force alone shipped a 63,000-gallon reserve on D-Day itself. By 12 June this reserve had increased, giving the British a stockpile of a million gallons. The estimated consumption of petrol in the early days of the campaign was of the order of 1,000 tons a day for motor transport alone, and for aviation a daily 700 tons was a moderate estimate.

To satisfy these immediate and massive demands on petrol by the Allies in France, four ship-to-shore pipelines were built and laid in each Task Force area for tankers to discharge their cargoes from offshore moorings direct to shore storage tanks, at Port-en-Bessin (British) and at Sainte-Honorine-des-Pertes (American), two miles further to the west. Codenamed 'Tombola', each pipeline enabled a large tanker to discharge 600 tons of fuel per hour. The system became fully operational by D+18. (Later, three more pipelines were hauled at the special request of the Americans at Fox Red Beach in the Omaha area.)

Tombola was relatively simple to set up. Fuel was pumped ashore from a tanker through an API (American Petroleum Institute industry standard) seamless steel pipe with an internal diameter of 6in. First, short lengths of this piping were jointed together by screwed sleeves to make 400ft sections, and then flanged together successively as they were hauled out to sea. On the sea end was a wooden sledge to ease the pipe over the sea bottom or, as used at Port-en-Bessin over rock, a special split buoy with a spherical end. (The foreshore there was not particularly suitable for the hauling operation, being very rocky, and the Great Storm of D+13 caused some delay.) A heavy flexible pipe was connected to the sea end once it had been fixed in-situ, the free end of the flexible pipe being sealed and buoyed. Moorings were constructed about the flexible pipe end so that a tanker lying to her own anchors ahead could have her stern held firmly over the pipe-end. She would haul up the flexible pipe and connect it to her own system and so pump her cargo of fuel ashore.

PLUTO (Pipeline Under the Ocean) did not begin pumping fuel across the Channel from its terminal on the Isle of Wight until 12 August – 10 weeks after the D-Day landings. In the final reckoning, PLUTO had virtually no effect at all on the supply of petrol to the Allies during the Normandy campaign.

RIGHT Operation Tombola initially pumped fuel from tankers several miles offshore. *(USNA)*

LEFT When Port-en-Bessin fell to the Allies, fuel was pumped directly from berthed tankers. From the harbour two 6in pipelines, with booster pumps, carried motor vehicle and aviation fuel to the US tank farm at Le Mont Cauvin, near Etréham, for use by British and American forces. *(USNA)*

LEFT In addition to the two pipelines from Port-en-Bessin there were two from Sainte-Honorine-des-Pertes. They joined up at Le Mont Cauvin where German prisoners filled jerrycans for use in the field. *(USNA)*

Mulberry B provided shelter during the storm to some 500 small craft and other vessels in the lee of the Corncobs. Without this protection the loss of small craft might have been catastrophic. In spite of the damage inflicted by the storm, unloading at Mulberry B continued.

The official report in 1946 by Sir Walter Monckton (Solicitor General in Churchill's 1945 caretaker government) into the part played by the Mulberry harbours in Overlord stated:

The difference between the effect of the storm on Mulberry A and Mulberry B is so great as to call for explanation. Some reasons are generally accepted:

■ Mulberry A took the gale square on the chin, whereas Mulberry B was struck a glancing blow. At Mulberry A the wave front was coming in parallel to the breakwater, while at Mulberry B it came in at an angle of 37° except for the period on D+13 when the wind was blowing from NNW.
■ Mulberry A had no natural shelter like the Calvados Shoal.
■ In Mulberry A the gaps in the Gooseberry breakwater were considerable, and the

individual blockships did not overlap sufficiently.
■ The bottom at Mulberry A had deeper sand and was accordingly more prone to scour.

As a result of the storm a major change was made in the construction programme for the two harbours. On 27 June, SCAEF decided that Mulberry A should not be rebuilt according to the original plan; there were to be no piers, but the No 2 Gooseberry was to be strengthened to provide shelter for small craft. This task was carried out using 22 Phoenix caissons and 12 Corncobs.

It was a different story for Mulberry B as SCAEF decreed that the harbour must last the winter; salvaged Whale bridges from Mulberry A were used to construct a second stores pier and an LST pier, as well as a shorter pier for barges. To bolster the harbour against autumn gales, the existing breakwater was double-banked with the addition of a second row of 40 Phoenixes. These extra precautions ensured the successful operation of Mulberry B up to its closure on 1 December 1944. In spite of the damage done by the storm, unloading never stopped completely.

What the Mulberries achieved

The use of ports captured after D-Day eventually affected the tonnage landed through the Mulberry harbours. Indeed, after the beginning of September, the capture of Dieppe, Ostend and later of Antwerp made it unnecessary to go on using Mulberry B. The statistics below and right reveal the effectiveness of the Mulberries in landing personnel, equipment and stores in the weeks following D-Day.

Note: Gooseberries 1 and 2 (American) were used for unloading purposes up to December 1944, more than six weeks after the British had ceased to use Gooseberries 4 and 5 for this purpose.

- Mulberry A had become Gooseberry 2 after the Great Storm.
- An average of 8,000 tons per day passed through Gooseberry 1, and some 15,000 tons a day through Gooseberry 2.
- Mulberry B (British) contained Gooseberry 3 at Gold Beach.
- Gooseberry 4 (Juno Beach) discharged an

average of 1,028 tons per day between D+6 and D+93 (7 September).
- Gooseberry 5 (Sword Beach).

The Mulberry harbours had many detractors who branded them a waste of resources that diverted huge amounts of materials, equipment and labour away from the war effort in general. Certainly, they were a voracious consumer of these commodities, but their contribution to the successful outcome of the Normandy invasion can be strongly argued, although it is outside the scope of this book.

Mulberry B

1. Vehicle and personnel discharge
(The cumulative British totals through all Continental ports, including Mulberry, are in brackets)

Period	Vehicles	Personnel
6 Jun–31 Oct	39,743 (236,358)	220,231 (964,703)

2. Stores
Total discharge up to 4 Sep: 517,844 tons (1,261,047 tons)
Maximum discharge on any one day: 11,491 tons (29 July)
(Source: SHAEF Handbook)

BELOW A scene of chaos on Omaha Beach. In the left foreground can be seen steel erection tanks used for assembling sections of Whale roadway; on the right centre are several LCT and sections of Bombardon washed up on the shore.
(USNA)

Chapter Four

Specialised armour

Hobart's Funnies and other oddities

The deployment of a diverse collection of specialised armoured vehicles and equipment in the assault phase of the Normandy landings enabled British and Canadian commanders to get their forces off the beaches quickly, saving vital time and lives. However, on Omaha Beach in the American sector it was a very different story.

OPPOSITE A Churchill AVRE Bobbin of 81st Assault Squadron can be seen in the surf towards the right in this photograph as soldiers of the 7th Green Howards make their way across the King Green sector on Gold Beach at about 8:30am on 6 June. *(IWM MH2021)*

The British decision to create a specialised armoured force was a result of the experiences of the raiding force at Dieppe, but it was not the true beginning of the development of British specialised armour, which had its birth in the First World War. When the war ended, so too did the need for such specialised armour and it was pretty much forgotten during the interwar period. When war broke out again in September 1939, ideas for specialised tanks were dusted down and new designs were put through their paces. Research and development was carried out to design weapons that could be used in unusual battle situations, like the Normandy beaches.

Some commentators have said that the British decision to centralise the development, organisation and use of specialised vehicles and equipment under a single command – the 79th Armoured Division – was much wiser than the decentralised approach taken by the Americans.

For many years it was an accepted fact that the heavy casualties suffered by the Americans on Omaha Beach on D-Day were attributable to the US Army's resistance to using specialised armour. However, closer examination of the documentary evidence by American historian and former US Army officer, Richard C. Anderson, led him to the convincing conclusion that 'there simply weren't sufficient British-made

Major General Sir Percy Cleghorn Stanley Hobart, KBE, CB, DSO, MC (1885–1957)

ABOVE Major General Percy Hobart, visionary armour tactician and thorn in the side of the military establishment. *(Author's collection)*

When invasion threatened in the summer of 1940, Major General Percy Hobart, one of Britain's greatest military tacticians and a commander of its armoured forces, was serving as a private soldier in the Local Defence Volunteers (precursor to the Home Guard) in the small Gloucestershire town of Chipping Campden. He had been forced into premature retirement earlier that year by conservative elements in the War Office. When this odd situation came to Churchill's attention, Hobart was quickly brought back to service at the premier's insistence and asked to form and train the new 11th Armoured Division.

Percy Hobart was born in India in 1885 and educated at Clifton College, Bristol, and at the Royal Military Academy, Woolwich, from where he graduated in 1904 and was commissioned into the Royal Engineers. He first saw service in India, and during the First World War in France and Mesopotamia. In 1923 he volunteered to be transferred to the Royal Tank Corps.

In 1934 Hobart was appointed to command the first permanent Tank Brigade. He soon showed himself as an outstanding leader and visionary when it came to the development of armoured forces. He broke with orthodoxy and moved away from the restricted role of close cooperation with infantry and instead developed the mobile technique for tanks.

Hobart's ideas on how armoured troops should be handled and what they could achieve on their own had been regarded with deep suspicion and dislike by his specialist vehicles to support the American operations, whether they were American- or British-manned. The American projects to develop similar equipment, begun in a similar timeframe as the British projects, simply were delayed too late for them to have been available in time. Those delays were, of course, unforeseen in mid-1943 when the projects were begun and in early 1944 when the equipment requests were made.'

He goes on to say that although an offer of special equipment developed by the 79th Armoured Division was made to the US Army 'it was not refused, and in fact a large number of types were asked for, but for various reasons were not supplied. The "refusal" of some of those items was for perfectly logical reasons – the difficulty associated with issuing brand-new, unique, and complicated items so close to the invasion date, as well as the mistaken belief that similar equipment, on standard American vehicles, would be supplied from the US.'

Hobart's Funnies

The strange-looking collection of specialised armour that made up the 79th Armoured Division acquired the nickname 'Hobart's Funnies' after its unorthodox but visionary

more conservative superiors at the War Office. In 1940 Sir Archibald Wavell sacked Hobart from his command of the Mobile Division, Egypt, which he had been sent to form after the Munich Crisis in 1938.

It was Hobart's 'unconventional' ideas about armoured warfare that had led to his demise. Field Marshal Montgomery (Hobart's brother-in-law) later said of this: 'He was a forward thinker and was a constant thorn in the side of senior officers less able than himself, particularly those who were inclined to plan the next war in terms of the last – an error common to military men. In the end this led to his downfall and he was retired from the Army when a major general.'

Hobart's successor, General O'Connor, who commanded the Western Desert Force in the first North African campaign in the winter of 1940–41, called Hobart's division 'the best trained division I have ever seen. (The Armoured Division, Egypt, was later to find fame as the 7th Armoured Division – the immortal Desert Rats.)

When the Dieppe operation in 1942 highlighted the need for specialised armour to spearhead an assault on a fortified coast, the CIGS Alan Brooke gave 'Old Hobo', as he was affectionately dubbed, the task of developing such specialised armour, and he was given the newly formed 79th Armoured Division to command and train. It became the first and only all-armoured formation in the British Army.

In the early stages the 79th possessed four main types of specialised armour: the DD (swimming tank), Crab (with flail for minesweeping), the AVRE (assault engineer tank) and the CDL (searchlight tank with dazzle

device). Later additions to Hobart's Funnies included the Crocodile (Churchill flame-throwing tank), Buffalo (armoured amphibious vehicle which, surprisingly, was not used on D-Day), and the Kangaroo (armoured personnel carrier).

Field Marshal Montgomery had this to say of the 79th Armoured Division: 'I have no hesitation in saying that he [Hobart] and his specialised division played a major part in the operations of 21 Army Group.'

Captain Basil Liddell Hart, the military historian and leading interwar theorist, commented on the 79th Armoured Division in his obituary of Hobart in *The Times*: '[it] played a vital part in the success of the Normandy landing, but a still bigger part in the final stage of the war. Quantitatively, it far exceeded the scale of any division. Qualitatively, it became the tactical key to victory.'

A fellow officer described 'Old Hobo' as 'a man of vision, whose capacity for work was apparently inexhaustible and whose enthusiasm, drive, and imagination were of a like nature. Above all he was both scrupulously fair and generous to his subordinates.'

commander, Major General Percy Hobart. When it came to the D-Day landings and the ensuing Battle of Normandy, Hobart's Funnies enabled battlefield commanders to get off the beaches more quickly and achieve their objectives with greater efficiency, saving vital time and lives in the process.

Six Royal Engineers (RE) assault squadrons with Armoured Vehicle Royal Engineers (AVRE), flails and armoured bulldozers, took part in the initial landings on the British beaches. The squadrons were subdivided into four troops with about 26 AVRE vehicles each, for obstacle clearance and mine gapping, with each troop or 'team' clearing a 'lane' up the beach. A 'team' usually comprised three or four AVREs and two flails. The leading AVRE in each team was equipped with a 'Roly-Poly' to lay a carpet from the ramp of the LCT to the dry beach, to cover the soft patches that were known to exist on the shore and could cause a tank to become bogged down. On the Canadian sector another assault squadron attacked, also with AVREs, flails and armoured bulldozers.

The versatile Churchill tank

The backbone of the 79th Armoured Division was the Churchill infantry tank. One of its greatest assets that made it ideal for service with the 79th was that it yielded readily to specialised modification. This was thanks largely to its roomy crew compartment, heavy armour, reliability and good cross-country performance.

Once tanks and vehicles had been landed on a beach they were certain to encounter ditches, tank traps and trenches that would need to be dealt with in order for the Allies to advance. This problem was addressed by the creation of a multi-purpose armoured vehicle, the Churchill AVRE, a specialised conversion of the Churchill Mk III and Mk IV gun tanks for assault engineer operations. The AVRE was to play a key part in the D-Day landings.

The multi-purpose AVRE hull for armoured engineer vehicles made it suitable for use in four main categories: demolition, mine clearance, ditch-crossers and road-makers. These included

RIGHT A 79th Armoured Division Churchill tank uses a Churchill ark to scale a sea wall during an exercise in the Saxmundham area of Suffolk on 11 March 1944. *(IWM H36593)*

the 290mm Petard spigot mortar that fired its 40lb 'Flying Dustbin' bomb over an effective range of 80yd to demolish 'stubborn' fortifications and buildings; the Churchill AVRE bridge-layer and ARK; the Churchill Bobbin; and the fascine (bundle of sticks) carrier for filling anti-tank ditches, trenches and other gap obstacles.

A total of 754 Churchill Mk IIIs and Mk IVs were converted to the AVRE specification during the Second World War, many being used by the 79th Armoured Division.

Churchill Petard spigot mortar

The 290mm spigot mortar that replaced the main gun armament on the Churchill Mk III or Mk IV was designed as a demolition weapon to destroy substantial structures like sea walls, bunkers, pillboxes and earthworks. It was also used to good effect in the Normandy bocage to blow gaps in the earthen banks of the sunken lanes. Most effective at a range of about 80yd, the Petard fired a 26lb finned explosive charge with an outer casing, giving a total weight of 40lb. The nickname 'Flying Dustbin' was because of the shape of the casing and some 10 rounds might be expected to breach a 6ft-thick concrete wall. The only drawback with the Petard was that the mortar barrel could usually only be reloaded from outside the tank turret by breaking the barrel like an air rifle.

Churchill ARK (Armoured Ramp Carrier)

With a crew of four the turret-less Churchill ARK carried hinged metal ramps at either end, with a single length mounted along the hull to form a mobile bridge. Developed in 1943 and based on the Mk III and Mk IV hull, the ARK was driven with the two hinged ramps in a vertical position until it reached an obstacle that needed bridging. It was then driven into a void in the ground or near a sea wall, where its two hinged ramps were lowered and positioned to create a bridge for following vehicles to drive across. If the void proved too deep for one ARK, another could be driven over the top of the first ramp carrier. There were two versions of the ARK, the Mk I and Mk II. Some 50 Churchill ARKs were built by REME workshops and the MG car company. It is believed that no Churchill ARKs were used on D-Day, but they were active in the Battle of Normandy and later.

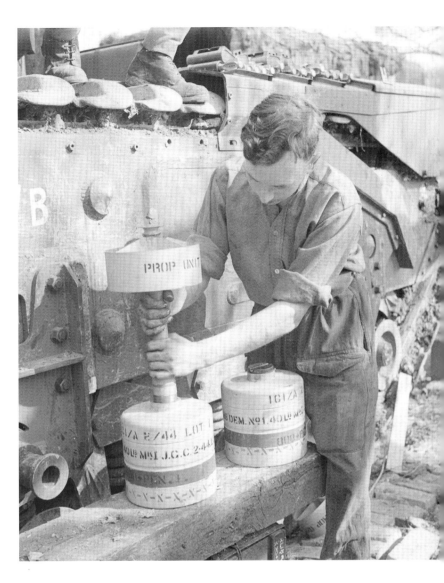

Churchill Bobbin

When the clandestine geological surveys of the Normandy coastline (see page 20) found that some of the proposed landing beaches had patches of soft blue clay (which would not support the weight of a tank), an innovative solution was reached in the design of the Churchill Bobbin using the Churchill Mk IV.

LEFT Churchill AVRE with spigot mortar. *(Author's collection)*

ABOVE 40lb bombs as used by the 29cm Petard spigot mortar on a Churchill AVRE of 79th Squadron, 5th Assault Regiment, Royal Engineers, under command of 3rd Infantry Division, 29 April 1944. *(IWM H38003)*

BELOW **Churchill Fascine.** *(IWM H37472)*

The Bobbin – or Churchill AVRE Carpetlayer Type C Mk II – was fitted at the front with two extending steel frame arms. Attached to them was a giant 'bobbin' on to which was wound a 10ft-wide carpet of scaffolding-reinforced hessian matting. When unrolled and laid over soft sand the carpet provided a firm base over which following vehicles (and the deploying tank itself) could pass without sinking into the ground. The carpet could also be used for laying over barbed-wire defences allowing them to be crossed.

Churchill SBG (Small Box Girder)

Carried on the front of a Churchill AVRE, this light but strong box-girder bridge saw important use on D-Day, allowing vehicles to cross over the sea walls from the beach, as well as any other obstacles that needed bridging. It could be deployed in 30 seconds to span a 30ft gap.

Churchill Fascine

To facilitate crossing a tank-proof ditch, a number of rolls of chestnut paling or brushwood were tied with wire rope into a bundle 8ft in diameter, 12 to 14ft wide and weighing about 4 tons, which was secured to the front of the AVRE. When a tank-proof ditch was reached the fascine was released from inside the tank, rolling off to make a causeway across which other vehicles could cross.

Churchill Crocodile

Arguably the most feared armoured fighting vehicle on the Normandy battlefield (although only a handful actually made it ashore on D-Day, none of which used their flamethrowers), the Churchill Crocodile instilled such fear in the Germans that a first ranging shot at a bunker might be enough to make its occupants surrender without a fight. One British soldier described how 'the effect of the fiendish screech and billowy roar of the flames on morale is terrific'.

A number of different designs were considered using the Valentine tank to mount a flamethrower, but the Department of Tank Design eventually opted for the Churchill as the basis for further development, being the infantry tank successor to the Valentine. At a demonstration of the Crocodile in 1943 Hobart saw great potential for the design and pushed the Ministry of Supply to consider production for his 79th Armoured Division, which it did.

Based on the Churchill Mk VII, the flamethrower was supplied as a kit that could be fitted in the field by a REME team. A detachable two-wheeled 6½-ton armoured trailer containing 400 Imperial gallons of fuel, dispensed by compressed nitrogen propellant, was attached to the rear of the tank by a three-way jettisonable armoured coupling. The trailer was connected to an armoured pipe that was fixed to the outer rear bulkhead of the tank and

ran along the underside to connect with the front hull-mounted flame projector in place of the BESA machine gun, and next to the driver's viewing slit.

The flamethrower fuel was actually a mixture of petrol, oil and rubber, making it a viscous, sticky substance that stuck to whatever it came into contact with. The fuel mixture burnt on water and could be used to set fire to woodland and buildings. The Crocodile could project a 'wet' burst of unlit fuel at a target, which would splash around corners inside trenches or in strong points, and could then be ignited with a second burst. Nowhere was safe from the intrusions of the Crocodile.

There was sufficient fuel in the trailer for 80 one-second bursts, or a series of bursts of longer duration, for which the crew needed first to prime the system. It took about 15 minutes before it reached the optimum operating pressure of about 600psi. When in use, the Crocodile burnt fuel at the rate of four gallons per second. Experience in Normandy showed that the trailer should not be pressurised more than 30 minutes before use in combat because of the risk of fuel leakages, and a drop below full operating pressure (like a slow puncture).

Some 800 flamethrower kits were produced, with 250 kept in reserve for operations against the Japanese.

RIGHT The connection from the fuel trailer (disconnected in this photograph) to the tank's hull. *(Author)*

Sherman tank conversions

The amphibious tank

The 'swimming tank' concept can be traced back to the years after the First World War when testing of amphibious tanks began, although not with much success. As the Second World War approached, swimming tanks were in vogue with the War Office, but testing the designs was beset with all manner of difficulties. However, the seeds were sown for what became known as the 'duplex drive', or DD amphibious tank, when the émigré Hungarian engineer Nicholas Straussler solved the problem with his ingenious folding canvas screen, which gave a tank buoyancy without adding much to its bulk.

Initial trials using the Tetrarch light tank proved successful, and production went ahead using the reliable Valentine that was already in service with the British Army. Fritton Lake near Great Yarmouth in Norfolk was requisitioned by the War Office for secret training activities by the 79th Armoured Division using the specially modified Valentine DDs.

The training of most DD tank crews (American, British and Canadian) comprised a two-week intensive course using the Valentine, where they learned vehicle maintenance, servicing and repair, navigation techniques and emergency escape procedures. This, the first training school for DD tanks, was known as Fritton Bridging Camp in an attempt to disguise its true purpose (as A Wing, Water Assault). B Wing was later formed at Stokes Bay near Gosport in Hampshire.

By 1944 the Valentine had been largely superseded by the newer American-built M4 Sherman, which by that time was in use by all the Allied armies. A major advantage enjoyed by the Sherman was that it could swim ashore with its gun pointing forward, ready to fight as soon as it landed.

Sherman DD – the swimming tank

Several hundred Shermans were modified in the UK to DD specification, giving them an amphibious capability so they could swim to the shore and provide vital armoured fire support to the first wave of troops landing on the enemy beaches.

A folding canvas screen was attached to a collapsible tubular metal frame that was secured to metal plates welded around the top of the tank's hull above the running gear. Two compressed-air cylinders stowed on the rear decking inflated 36 rubber tubes or pillars, raising the screen and its frame, which were then secured by jointed metal bracing stays that

RIGHT A Sherman DD with its screen lowered, screen locking struts on the engine deck and the twin DD propellers in the raised (disconnected) position. *(Tank Museum)*

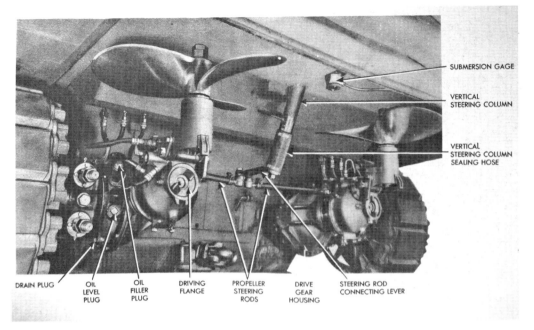

DRAIN PLUG   OIL LEVEL PLUG   OIL FILLER PLUG   DRIVING FLANGE   PROPELLER STEERING RODS   DRIVE GEAR HOUSING   STEERING ROD CONNECTING LEVER

SUBMERSION GAGE

VERTICAL STEERING COLUMN

VERTICAL STEERING COLUMN SEALING HOSE

LEFT Schematic of the twin propellers in raised position. *(Tank Museum)*

BELOW A three-quarter rear view of a Sherman DD with its screen fully raised and its propellers engaged. *(Tank Museum)*

BOTTOM A Sherman DD with its screen and propellers raised enters the waters of the River Rhine in March 1945. *(Author's collection)*

were manually locked in place. Once ashore the bracing stays were 'broken' and the screen could be folded down like a concertina.

A tarred sealing compound was used to waterproof the hull of the tank, and to cover those parts of the flotation screen where canvas and metal were joined together, particularly where bolts passed through the canvas.

The rubberised waterproofed canvas screen comprised three circular sections stacked one on top of the other like rings: the lowest was three-ply thick; the middle two-ply and the top a single thickness. It took eight minutes to inflate the screen, raising it to a height of 7ft, whereby the 30-ton tank displaced enough water to make it float. A hand-operated bilge pump was carried on the hull inside the canvas screen. With about 3ft of freeboard (that is, the height of the screen above the water line) when afloat, the DD tank could easily pass for a canvas boat, serving to confuse enemy observers.

Because of the design of the transmission in the Sherman it was not possible to take a shaft directly from the gearbox to drive the rear-mounted twin propellers, so a secondary drive system using an extra sprocket took power off the rear idler, which enabled the tank to swim at four knots (4.6mph) from a landing craft to the beach. The propellers were controlled by the driver via a hydraulic system. 'DD', or 'duplex drive', was the name given to this secondary

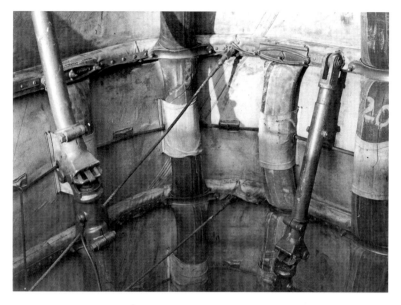

ABOVE RIGHT A rear view of the Sherman DD. *(Tank Museum)*

ABOVE Detail of the air column rubber tubes and jointed metal bracing stays. The air required to inflate the rubber tubes was provided by two air cylinders containing air compressed to 1,800lb/sq in when full. *(Tank Museum)*

drive system. Directional control was achieved by use of a small rudder and by swivelling the propellers on a horizontal axis. The driver needed a periscope to see where he was going, but was aided by the commander who could communicate with him by radio and/or steer the tank by means of a large tiller.

Once ashore, the canvas screen was expendable and it was assumed that tank crews would discard it as soon as the situation presented itself. However, some armoured units kept their flotation equipment attached and it was used in later amphibious operations, like the Rhine crossing in March 1945.

On D-Day, Sherman DDs partially equipped eight tank battalions of the American, British and Canadian forces. They were carried across the Channel in Landing Craft, Tank (LCT) and

RIGHT A Sherman DD tank, with its screens removed, passes through Douet near Pegasus Bridge as engineers work to clear the debris, 25 June 1944. *(USNA)*

LEFT Detail view of a Sherman DD's twin propellers in the engaged position. The furthermost propeller demonstrates its castoring capability to assist in steering the tank in water. *(Author)*

typically were launched at around two miles from the shore, from where they could swim ashore. DDs were used with success on all the Allied beaches except for Omaha, where 27 out of the 29 DDs launched into the sea sank, although many of the crews were rescued. The remaining tanks (mainly conventional Shermans and a handful of DDs) were eventually landed direct on to the beach.

Deep-wading Shermans and Churchills

Another solution to launching tanks into water for amphibious assault was to attach what was called deep-wading gear, which allowed a tank to drive along the seabed (rather than swim like the DD) while partially submerged. This modification had been used with success in Sicily in July 1943, and in subsequent

LEFT The 746th Tank Battalion's Sherman, *Hurricane*, fitted with wading trunks and waterproofing to its gun mantlet, disembarks on Utah Beach on D-Day. *(USNA)*

Mine-killing Sherman – the Crab flail tank

During the Second World War, both sides laid millions of anti-tank and anti-personnel mines, which posed a serious threat to advancing vehicles and troops. Many ideas for dealing with the mine menace were considered and tested, but most failed to get past the prototype stage and were abandoned. However, the flail was one solution that made it beyond the prototype stage, to become arguably the most successful of all the mine clearance devices to be developed in the war.

Early development work on the flail concept was undertaken by Captain Norman Berry and Major L.A. Girling in Egypt during 1942, and it resulted in a Matilda tank being fitted with a revolving drum rotor carrying 24 mine flails (heavy metal chains). This was mounted on two projecting arms at the front of the tank. It became known as the Matilda Scorpion, and at the second Battle of El Alamein in October 1942, 25 Scorpions were used with modest success to clear German minefields, although breakdowns were frequent owing to the rushed development of the flail system, and much of

ABOVE Sherman V Crab Mk II minesweeping flail tank in action, used to clear already identified minefields.
(IWM H38079)

BELOW Sherman V Crab Mk II equipment stowage, near side front view.
(Tank Museum)

amphibious operations in Italy. The hull was waterproofed, and rectangular metal trunks were fitted to the decking, one over the engine air intake and the other over the exhaust, extending upwards from the engine deck to above the turret top to prevent water ingress. Once ashore, these devices were unbolted and discarded.

Churchill tanks with deep-wading trunks had taken part in the Dieppe raid and they were used again on D-Day. Deep-wading gear was also used by the Allies on M-10 tank destroyers, light tanks, trucks, Universal Carriers, and Jeeps.

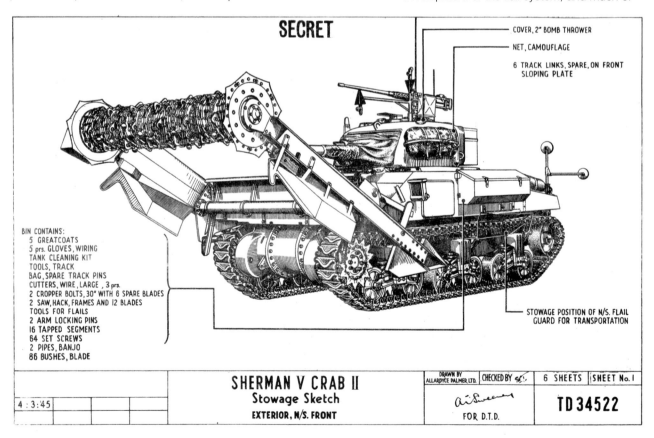

SECRET

COVER, 2" BOMB THROWER

NET, CAMOUFLAGE

6 TRACK LINKS, SPARE, ON FRONT SLOPING PLATE

BIN CONTAINS:
5 GREATCOATS
5 prs. GLOVES, WIRING
TANK CLEANING KIT
TOOLS, TRACK
BAG, SPARE TRACK PINS
CUTTERS, WIRE, LARGE, 3 prs.
2 CROPPER BOLTS, 30" WITH 6 SPARE BLADES
2 SAW, HACK, FRAMES AND 12 BLADES
TOOLS FOR FLAILS
2 ARM LOCKING PINS
16 TAPPED SEGMENTS
64 SET SCREWS
2 PIPES, BANJO
86 BUSHES, BLADE

STOWAGE POSITION OF N/S. FLAIL GUARD FOR TRANSPORTATION

SHERMAN V CRAB II
Stowage Sketch
EXTERIOR, N/S. FRONT

4 : 3 : 45

DRAWN BY ALLARDYCE PALMER LTD. CHECKED BY

FOR D.T.D.

6 SHEETS   SHEET No. 1

TD 34522

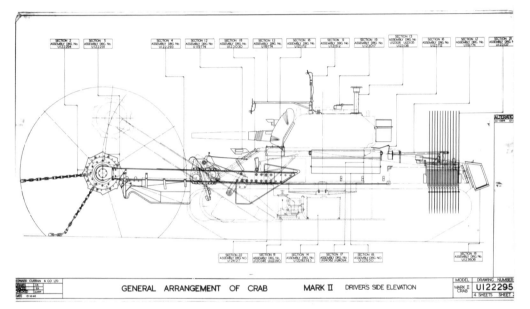

ABOVE This disabled
Sherman Crab of the
Westminster Dragoons
on Sword Beach
on 7 June shows to
advantage its rotating
drum and flails.
(IWM B5141)

LEFT General
arrangement of the
Crab Mk II, driver's
side elevation.
(Tank Museum)

the mine clearing had to be done the traditional
way – by hand. Further versions of the Scorpion
were developed, based on the M3 Grant tank,
with a small number being used in the North
African campaign and later during the Allied
invasion of Sicily.

Meanwhile, Captain Abraham du Toit, a
South African, had been sent to Britain to work
alongside AEC Ltd and develop an effective
mine-clearing device. He arrived at the same
solution as his contemporaries in Egypt, using
threshing chains attached to a rotor mounted
on a tank. Initial trials involved a Matilda Mk II
tank called the Baron Mk I, followed by the
Baron Mk II, both of which were tested during
1942–43. Both the Scorpion and the Baron
used externally mounted auxiliary motors to
drive the rotor, which made them too wide to
cross a Bailey Bridge or to be carried on rail
flats. Some 60 Barons were eventually built,
but they were used mainly for training and

demonstrations in the UK.

Development of the flail concept continued,
with further designs based on the M4A4
Sherman, the Sherman Scorpion Mk IV and Mk V,
which used chains on the front sprockets of
the tank to drive the flail drum. By mid-1943
significant progress had been made with the
design and an improved version of the device,
still using the Sherman, evolved into the
Sherman Crab. The first user trials of the Crab
began in October 1943, when it was found that
the design was particularly effective at cutting
barbed wire as well as in its primary role of
mine clearance.

The Sherman Crab had its first baptism
of fire on D-Day when operated by the 30th
Armoured Brigade in the famous 79th Armoured
Division. There were three flail regiments in
the 30th Armoured Brigade – the Westminster
Dragoons, the 22nd Dragoons and the 1st
Lothian and Border Horse.

ABOVE The dozer
blade fitted to the
Sherman tank dozer
pivoted on brackets
fixed to the middle
bogie assemblies of
the tank. A hydraulic
jack powered by an
oil pump driven off
the propeller shaft
was mounted on the
outside front of the
tank to raise and
lower the blade. This
Sherman dozer is
pictured in Normandy
on 4 July 1944.
(IWM B6371)

Sherman tank dozer

The versatile Sherman tank underwent a
further specialised modification in 1943 with
the addition of the hydraulically operated
LaPlante-Choate dozer blade from a
Caterpillar D8 tracked tractor. In due course
the M1 dozer blade was adapted to fit all
Shermans using the Vertical Volute Spring
System (VVSS) suspension; the M1A1 blade
was fitted to the M4A3 Sherman model
onwards, which used the less common
Horizontal Volute Spring System (HVSS).
The Sherman tank dozer retained its main
armament and fighting capability, and in an

emergency its driver could jettison the blade
attachment within 10 seconds.

Armour protection kept tank dozer crews
safe and the dozer installation itself was simple
and reliable in use, with minimal maintenance
requirements. However, there was a drawback
to its use: the dozer blade overloaded the tank's
front suspension system, increasing bogie tyre
failures and suspension maintenance demands.

On D-Day tank dozers were used to destroy
German beach obstacles, which included
pushing barriers and mines out of the way at
low tide. The Americans planned to use tank
dozers on Omaha and Utah Beaches with
engineer special assault gapping teams during
the assault phase. On Omaha 16 M4 tank
dozers were to have landed, but only six finally
made it ashore. Engineers used the remaining
dozers to shove the obstacles aside once
any attached explosives had been dealt with.
Eventually the Germans knocked out all but one
of the dozers. Later in the morning, armoured
bulldozers arrived to expand the exit routes off
the beach and to build roads across the dunes.

The assault on Utah Beach was less
troublesome than that on Omaha, which meant
the two tank dozers that landed could work
faster to clear beach defences and open up
exit routes.

RIGHT The Sherman
'Rhino' or 'Culin
cutter' was a simple
but effective way of
breaking through the
thick earthen banks of
the Normandy bocage.
(USNA)

Hedgerow-buster – the 'Rhino' or 'Culin cutter'

A simple but ingenious piece of equipment for breaking through the steep earthen banks topped with dense hedgerows of the Normandy bocage was devised by Sgt Curtis G. Culin of the US 102nd Cavalry Reconnaissance Squadron. Using salvaged steel plate from German beach defences, sections of rail discarded track and steel pipe, Culin fabricated sets of steel prongs that were welded to the lower front hulls of Sherman tanks, projecting approximately 2ft forward. These Shermans became known variously as 'Rhinos' and 'Culin cutters' (after their inventor). The 'Rhino' would be run into an earthen bank and use its steel prongs to loosen the dirt and roots, thus enabling it to crash through to the next field. In use from July 1944, the Rhino conversion was simple, reliable and effective.

Armoured bulldozers

It was the British who developed armoured bulldozers in the Second World War. A standard Caterpillar D8 bulldozer, fitted with armour plate to protect the driver and the engine, was produced in large numbers in the run-up to D-Day. Once ashore the tasks of the

armoured bulldozer were clearing the invasion beaches of obstacles, building roads through the dunes and making the exit roads off the beaches accessible to vehicles by clearing away rubble and filling in bomb craters.

BELOW A scene of intense activity on Nan Green Beach in the Juno sector. An armoured Caterpillar D8 bulldozer can be seen on the right of the picture, fitted with a LaPlante-Choate dozer blade. In the centre is a Sherman III gun tank. *(National Archives of Canada)*

ABOVE The Sherman Beach Armoured Recovery Vehicle (BARV) was not operated by the 79th Armoured Division, although it is sometimes referred to as one of Hobart's Funnies. Some 60 BARVs were used on the Normandy beaches to remove broken-down vehicles, those that had foundered in the surf and were blocking access to the beaches, and to refloat small landing craft stranded on the beach. Developed and operated by the Royal Electrical and Mechanical Engineers, the BARV was based on the Sherman M4A2 chassis. Its turret was removed and replaced with a tall armoured superstructure, its hull was waterproofed and deep-wading trunking was fitted, enabling the tank to operate in up to 9ft of water. Unlike other Sherman models, this was powered by a diesel engine. *(IWM B5578)*

Chapter Five

Assault from the skies

Airborne operations

In the early hours of 6 June audacious landings by British and American paratroopers and glider-borne infantry seized key positions on the flanks of the invasion area. These airborne assaults at Pegasus Bridge and Sainte-Mère-Église were made possible through the use of specialised assault gliders and paratroop aircraft guided by radio homing devices.

OPPOSITE Late evening on 5 June, British Pathfinder stick commanders of 22 Independent Parachute Company synchronise their watches against the backdrop of a 295 Squadron Albemarle paratroop transport aircraft. Left to right: Lieutenants Bobby de Lautour, Don Wells, John Vischer and Bob Midwood. De Lautour was injured during the fighting and died of his wounds on 20 June. *(IWM H39070)*

RIGHT Modern-day
paratroopers drop
from a C-130 Hercules
over Normandy on
the 60th anniversary
of the D-Day landings
in 2004. (US Dept of
Defense)

ABOVE Bellerophon
riding Pegasus. The
flying horse was the
emblem of British
airborne forces in the
Second World War.
(Author)

ABOVE The Parachute
Regiment was
among the youngest
regiments in the
British Army when
it was formed on
1 August 1942.
(Author)

For millennia, armies had attacked overland or from the sea, but with the advent of the aeroplane came the possibility of parachuting men into the battle area from above. Parachute troops (paratroopers) could avoid fixed defensive fortifications and force a defender to spread his defences to cover other areas of territory normally protected by geographical features.

As well as their use as shock troops, paratroopers could be deployed to establish an airhead for landing other units. In the case of the assault on Normandy in 1944, special Pathfinder teams of paratroops were used to secure and mark out the drop zones (DZ) and landing zones (LZ) for further paratroops and assault glider-borne infantry (another method of delivering troops from the air). However, the limited capacity of transport aircraft of the period (like the Douglas DC-3/C-47 or the German Junkers Ju 52) meant paratroops rarely, if ever, jumped in groups larger than 20 from one aircraft.

Germany made widespread use of paratroopers in the Second World War. It made the first airborne assault of the war when, on 9 April 1940, its *Fallschirmjäger* (paratrooper) units spearheaded the invasion of Denmark as part of Operation *Weserübung*. Inspired in part by the successes of the German *Fallschirmjäger*, Britain formed its own airborne forces, which included the Parachute Regiment,

ABOVE The US 82nd
Airborne Division was
formed out of the 82nd
Infantry Division, which
acquired its nickname
'All American' during
the First World War
when it was discovered
that the division
contained draftees
from the 48 US states
of the union that
existed at the time. It
was redesignated as
an airborne division
in 1942. This is the
division's shoulder
sleeve insignia, the
'AA' standing for 'All
American'. (Author)

ABOVE It was during
the 1920s that the
'Screaming Eagle'
became associated
with the US 101st
Division, a US Army
reserve formation
based in Milwaukee,
Wisconsin, as
successor to the
traditions of the 8th
Wisconsin Volunteer
Infantry Regiment of
the American Civil
War. The 101st was
reorganised as an
airborne division in
1942. (Author)

Air Landing Regiments, and the Glider Pilot Regiment. Its first airborne assault took place on 10 February 1941 when Special Air Service (SAS) troops parachuted into southern Italy to blow up an aqueduct at Calitri in the daring raid codenamed Operation Colossus.

In the USA the concept of airborne units was not popular with military commanders, but under the patronage of President Franklin D. Roosevelt the first paratrooper platoon was formed in 1940, which in time led to the creation of the US Army's Airborne Command. The first US Army combat jump was near Oran in Algeria on 8 November 1942 by elements of the 509th Parachute Infantry Regiment.

In early 1942 the US War Department decided to form two airborne divisions. The first to be chosen was the 82nd Infantry Division, which became the 82nd Airborne Division under the command of Major General Omar N. Bradley, with Brigadier General Matthew B. Ridgway as his second-in-command. (Both men went on to achieve high office in the US Army and were to play key roles in Operation Overlord.) On 15 August the 82nd was split and two airborne divisions were formed – the 82nd and the 101st. The organisational structure of each division was initially to be two glider infantry regiments and one parachute regiment (although this was to be very different on D-Day), plus the usual divisional support troops (artillery, signals, medical, engineer and transport). Command of the 101st was given to Major General William C. Lee.

Assault gliders

In the Second World War the assault glider was used to deliver troops and their equipment to the battlefield, a job that is undertaken in the 21st century by the helicopter. A rapid delivery technique was developed so that the time taken to land by a glider after release from the tug aircraft was minimal, which meant it was exposed to enemy ground fire for as short a time as possible.

Germany was the first nation to use gliders to great effect during the invasion of Belgium in 1940. On 10 May of that year, 78 DFS230 glider-borne paratroopers of the 7th *Flieger* Division (which later became the 1st

Fallschirmjäger Division) landed on the roof of the reputedly impregnable Eben-Emael fortress on the Belgian–Dutch border, near the Albert Canal. Armed with high explosives, they achieved complete surprise, taking the fortress and neutralising much of its defensive armament within minutes.

Prime Minister Winston Churchill set in motion the formation and training of a British airborne force within weeks of the fall of France in June. By September a Glider Training Squadron had been formed, but heated discussion followed as to which service the men who would fly the gliders would belong, the RAF or the Army. It was finally decided that the glider pilots would belong to the Army, but would be trained to fly by the RAF. It was in the same month that the glider that was to form the backbone of Britain's airborne forces, the Airspeed Horsa, made its first flight.

LEFT British Glider Pilot Regiment pilot's wings. Volunteers for glider pilot training were drawn from both the officer corps and the ranks. They sat a joint Air Force/ Army selection board and, if accepted, underwent six weeks' basic training at the GPR Depot, Tilshead, on Salisbury Plain, followed by 30 weeks of flying training before they were awarded their wings. As well as being able to fly a glider, all pilots were expected to reach a particular standard as a soldier. *(Author)*

LEFT A glider pilot of D Squadron Glider Pilot Regiment (GPR) stands beside his Horsa at RAF Keevil in Wiltshire. *(Author's collection)*

RIGHT CG-4A pilot in the cockpit of his glider. Note the tubular frame construction of the aircraft. *(USNA)*

BELOW USAAF glider pilot's silver wings. Most USAAF glider pilots came from enlisted ranks and all were volunteers. Upon completion of their flying training, enlisted men would be promoted to staff sergeant (or would retain their existing grade if higher) while officers would train in grade. After 21 November 1942, all enlisted graduates were appointed as flight officers, equal to the existing rank of warrant officer Junior Grade (WO1) upon completion of their advanced glider training. *(Author's collection)*

On 10 October 1941, the 1st Air Landing Brigade was formed under the command of Major General F.A.M. 'Boy' Browning, and was made up from the 1st Royal Ulster Rifles, 2nd Oxfordshire and Buckinghamshire Light Infantry, 2nd South Staffords and the 1st Border Regiment. The 1st Air Landing Brigade became part of the British 1st Airborne Division and was carried into battle by gliders on the eve of D-Day, and after.

With the likelihood of more than half of Britain's airborne forces being carried by glider, the Glider Pilot Regiment (GPR) was formed on 21 December 1941, its pilots being raised from Army volunteers. The GPR was to be made up from two battalions of six companies of sergeant pilots, who were also trained as infantry soldiers to fight on the ground after landing. (This was a reflection more of necessity than desire, however, since the main idea was to evacuate these highly trained specialist personnel back to the UK as soon as possible after the landings.)

On 25 February 1941, the commanding officer of the US Army Air Corps, General Henry 'Hap' Arnold, ordered a report on the use of gliders for US forces. Within a week Arnold had requested USAAC design engineers based at Wright Field, Dayton, Ohio, to design gliders to carry 12 to 15 troops with their equipment, or a load of military supplies. With the entry of the USA into the war in December 1941, its glider-building and glider pilot training programmes were speeded up. Glider infantry regiments (GIR) were formed under the command of one or other of the two airborne divisions. Their pilots were members of the GIR and were given the rank of flight officers (warrant officers), with the opportunity of becoming a commissioned officer. They differed from their British GPR counterparts by not being in a separate glider pilots' regiment, and they were not trained to fight on the ground after landing, their primary responsibility being to fly the glider to the LZ, land it and unload it.

Airspeed Horsa

It took just 10 months for the Airspeed Horsa glider to go from the drawing board to its first flight on 12 September 1941. Designed to Ministry of Aircraft Production Specification X26/40 for a military assault glider of wooden construction capable of carrying 25 troops and two pilots, aircraft builder Airspeed Ltd (a subsidiary of the de Havilland Aircraft Co.) carried out design work on the AS51 Horsa Mk I at de Havilland's Salisbury Hall design offices under the direction of A. Hessell Tiltman (co-founder of Airspeed) and designer A.E. Ellison.

Following its completion, most of the design team were moved to Portsmouth where they worked on the AS58 Horsa Mk II, which featured a hinged nose to allow vehicles or artillery pieces to be carried.

Towed aloft from Fairey's Great West Aerodrome at Hounslow (now engulfed by Heathrow Airport) on 12 September 1941 by an Armstrong Whitworth Whitley, the Horsa prototype, DG597, was flown by Airspeed's Chief Test Pilot George Errington (whose career was to end tragically in June 1966, at the age of 64, when he was killed co-piloting a Hawker Siddeley Trident, G-ARPY, on a test-flight).

The Horsa was characterised by its high wing and its long cylindrical fuselage, a considerable portion of which was forward of the main planes. A tricycle undercarriage was fitted (the first British glider to have one) that had a castoring nose-wheel in a fork mounting

RIGHT An advertisement in a wartime issue of *Flight* magazine for the 'Horsa, British standard towed transport'. *(Author's collection)*

on a short vertical shock absorber leg. The two main wheels were mounted on a split axle, with strut shock absorbers on each wheel attached to the wing near the roots.

For actual operational landings the main undercarriage could be jettisoned after take-off, and the landing was made on the nose-wheel and a sprung central skid. The glider heeled over until one of the wingtip skids touched down. However, the main undercarriage was invariably retained since it shortened the stopping distance on touchdown and gave the

BELOW The high 'slab' wing and barrel-shaped fuselage of the Horsa can be appreciated in this picture. *(USNA)*

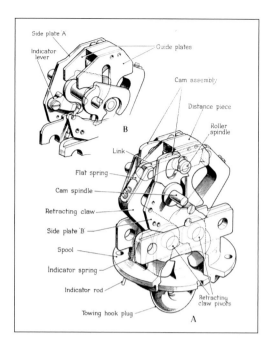

unloaded by either detonating a small explosive charge to blow off the rear fuselage, or if time and circumstances permitted the tail section could be unbolted from the main fuselage with a spanner.

Flown mainly by NCO pilots from the Army's Glider Pilot Regiment and RAF glider pilots, Horsas were used for transporting the men and equipment of the Air Landing Brigades in the British 1st and 6th Airborne Divisions.

Less than two years after its maiden flight, the Horsa was put to the test in its first significant operation during the invasion of Sicily in July 1943, and later with success on D-Day (6 June 1944); in the invasion of southern France (Operation Dragoon in August); at Arnhem (Operation Market Garden in September); and during the Rhine crossing (Operation Varsity) in March 1945. Some 400 Horsas were also used by US airborne forces in a sort of reverse Lend-Lease arrangement (although the Americans claimed not to like them).

Horsa production was subcontracted to a number of companies, particularly those in the furniture, cabinet and piano-making trades.

pilots better directional control of the glider on a congested landing zone.

Immediately behind the cockpit compartment in the nose, a door hinged down to provide a sloping gangway for troops to enter or leave the aircraft. On the battlefield, heavy equipment was

RIGHT With assistance from five pairs of helping hands, a Jeep is driven up wheel tracks into the loading hatch of a British Horsa during an exercise on 22 April 1944. *(IWM H37692)*

LEFT The rear fuselage of the Horsa could be unbolted to offload vehicle cargo. Alternatively, when in a hurry a small explosive charge fitted to the rear fuselage joint could be detonated to blow off the tail. 'Usually the nose end wouldn't budge and the tail end wouldn't blow off with the explosives,' recalled one British soldier, 'so we just ended up cutting our way out using axes.' (USNA)

Only a handful of the 3,799 Horsas eventually built were actually made by Airspeed. Several hundred were manufactured by the Austin Motor Company at Cowley, but the great majority were constructed by Harris Lebus, the furniture makers, at their Ferry Lane factory in Tottenham, North London, aided by various other subcontractors.

For example, Harris Lebus machined all wooden parts to practically the finished size before sending them to the parts store. From there they were reissued to the sub-assembly

LEFT The view from a Horsa cockpit with the Stirling tug visible towards the top of the picture. (Author's collection)

departments for building up into larger units. These in turn were sent to store and supplied as needed to the assembly departments concerned.

Arrester chutes were fitted to a number of Horsas but they were ineffective in many cases because the glider was flying too slowly.

Specification –
AIRSPEED AS51 HORSA MK I

Crew	2
Capacity	25 troops; or six-pounder anti-tank gun and crew, plus towing Jeep and ammunition
Length	67ft 0in
Wingspan	88ft 0in
Height	19ft 6in
Wing area	1,104sq ft
Empty weight	8,370lb
Loaded weight	15,500lb
Maximum speed	150mph on tow; 100mph gliding
Wing loading	14.0lb/ft

Production:

Airspeed prototypes	7	
Horsa Mk I	2,231	(Airspeed 470; Austin 300; Harris Lebus 1,461)
Horsa Mk II	1,561	(Airspeed 225; Austin 65; Harris Lebus 1,271)
Total:	3,799	

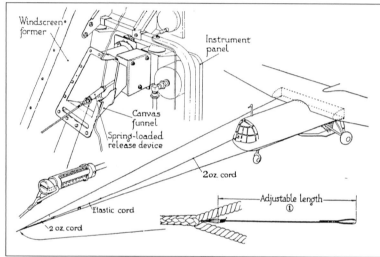

TOP Horsa fuselage sections under construction in the Austin factory at Cowley, Oxford. *(Author's collection)*

ABOVE 'Angle of dangle' attachment schematic. *(Assault Glider Trust)*

RIGHT Low tow – a Horsa behind a Stirling. *(Author's collection)*

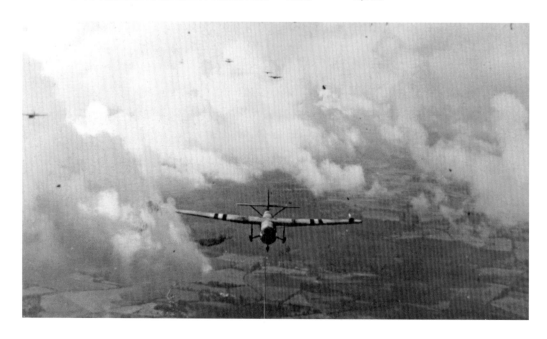

Wonder glue and paper floors – urea-formaldehyde adhesives

With the outbreak of war and an envisaged shortage of aluminium, the British aircraft industry reverted to the use of wood. The fund of knowledge acquired over generations by the UK's boat-builders, in the preparation of laminated hulls and general bonding (or gluing) applications, was seized upon by the Ministry of Aircraft Production, which was eager to make use of it for building wooden gliders.

The big breakthrough had come during the 1930s with the development of synthetic urea-formaldehyde (U/F) adhesives. An organic chemist by the name of Norman de Bruyne developed the Aerolite range of U/F wood glues, which were strong, quick-drying and stable. They quickly replaced the old-fashioned glues based on natural products, which were subject to degradation and therefore unreliable. The new Aerolite materials were approved by the Ministry of Aircraft Production for use in aircraft building, and were subsequently used extensively in the manufacture of wood laminates and in aircraft construction (the Horsa glider and the Mosquito fighter-bomber, for example). In the early 1940s, de Bruyne developed a strip heating process that reduced curing (hardening) times with Aerolite adhesives from several hours to a matter of minutes.

The benefits that strong, quick-drying U/F brought to aircraft mass production, and the Horsa glider in particular, are plain to see. The bonding techniques developed for the aircraft industry were taken up after the war by the automotive industry.

Papreg

One of the USA's most promising wartime developments in the general field of plastics was by the Forest Products Laboratory, which devised a paper-based laminated plastic known as papreg. This material was prepared by impregnating special paper with phenolic resins, then moulding the paper sheets into a laminated plastic. Papreg attracted the interest of aircraft manufacturers as well as paper manufacturers, because it was found to be half as heavy as aluminium and yet capable of developing a tensile strength of 35,000 to 50,000psi, which was comparable to that of certain aluminium alloys on a relative weight basis. Apart from its high tensile strength, papreg also offered exceptional dimensional stability, low abrasiveness, and high impact-resistance.

During the Second World War, papreg was used in the manufacture of floors for more than 156 assault gliders, which were delivered to the Northwestern Aeronautical Corporation in Minneapolis/St Paul, Minnesota, but it is not clear if any of these were used on D-Day.

Paper laminates were set to become a critical component in aircraft construction, which included assault gliders and the de Havilland Mosquito.

General Aircraft GAL 49 Hamilcar

Designed by General Aircraft's Chief Designer, F.F. Crocombe, to Air Ministry Specification X27/40, the troop and heavy-equipment-carrying Hamilcar was the largest and heaviest Allied glider of the war, and the largest wooden aircraft ever constructed. It was the first Allied glider capable of carrying a seven-ton tank and possible loads included a Tetrarch Mk IV, US Locust tank, two Bren Carriers or scout cars, or a mobile Bofors gun, and other equipment such as bulldozers and Bailey bridge components.

A half-scale trial model was constructed, which was followed by a full-size prototype that flew on 27 March 1942. A total of 390 Hamilcar

LEFT 'Britain's latest secret weapon' proclaims this wartime advertisement for the General Aircraft Hamilcar glider. *(Author's collection)*

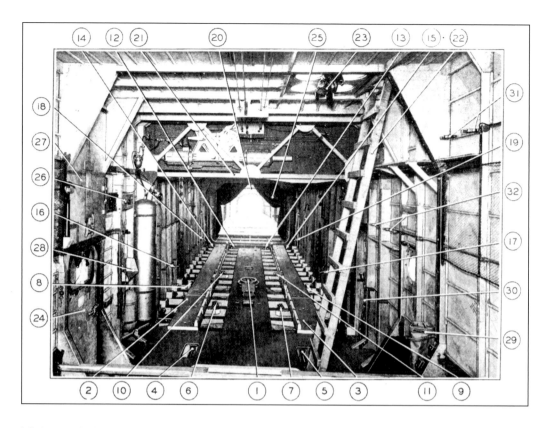

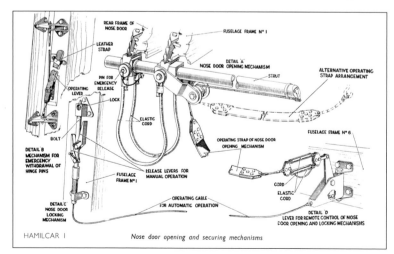

Mk Is were built, all but the first example by
a consortium of 22 furniture manufacturers.
A later version, the Mk X, was fitted with two
965hp Bristol Mercury radial engines.

The first 22 Hamilcars were built by General
Aircraft, with the remainder constructed by a large
team of subcontractors in what was termed the
Hamilcar Production Group, under the control
of the Birmingham Railway Carriage & Wagon
Company at Smethwick, which included the
Co-operative Wholesale Society and AC Cars.

Like the Horsa, the Hamilcar was a high-
wing aircraft, but it featured a hinged nose
to allow straight-in loading to its large 25ft
6in-long, 8ft 0in-wide and 7ft 6in-high freight
compartment, in theory enabling an armoured
vehicle to be in action in as little as 15 seconds
after the aircraft had come to rest. To facilitate
this, the vehicle engine was started up in the air
before landing, with temporary extension pipes
attached to the vehicle's exhausts to carry away
the fumes to the outside of the glider. These
pipes disengaged as the vehicle drove forward
to leave the freight compartment, at the same
time operating a device that freed the nose-
door lock and automatically opened the door.

It was originally intended for the Hamilcar to
make skid landings when used operationally,
and for take-off it was fitted with a special
¾-ton chassis that could be dropped by
parachute once airborne. However, with the
evolution in airborne landings strategy as the
war progressed, changes in landing techniques
were introduced. Because available space on
glider landing zones was usually very restricted,
and in order that they could be used by as
many gliders as possible, it was necessary to
keep them clear. Therefore it was decided that
gliders should land on their normal, wheeled,

undercarriage chassis, and use their speed combined with differential wheel braking to steer themselves clear of the landing strip. In this context a noteworthy feature of the Hamilcar's design was the large and powerful pneumatically operated wing flaps, which enabled the pilot to control the angle of glide and to make a landing in a confined space. As soon as a glider came to rest, high-pressure oil in the chassis shock absorbers was released, causing them to telescope and allowing the glider to sink on to its skids for the vehicle inside to drive out without needing ramps.

Because of the Hamilcar's greater size and weight, the Handley Page Halifax was found to be the only tug aircraft capable of towing the glider, although on occasion the Short Stirling was also used.

The Hamilcar's first operation was by the British during the D-Day landings on 6 June 1944, when some 70 took part. It also saw action at Arnhem in September 1944 (28 aircraft) and on the Rhine crossing the following spring.

ABOVE The Halifax was the only aircraft capable of towing the heavy Hamilcar. *(IWM EMOS1357)*

LEFT Hamilcar gliders of 6th Air Landing Brigade, flown by pilots of C Squadron GPR, descend on DZ 'N' on 6 June, carrying Tetrarch tanks of the 6th Airborne Division Armoured Reconnaissance Regiment. *(IWM B5198)*

Specification –
GENERAL AIRCRAFT LTD GAL 49 HAMILCAR MK I

Crew	2
Capacity	17,500lb maximum payload; 60 fully equipped troops or a 7-ton Tetrarch tank
Length	68ft 1in
Wingspan	110ft 0in
Height	20ft 3in
Empty weight	19,500lb
Max loaded weight	37,000lb

Performance:

Max towing speed	150mph
Max diving speed	187mph
Stall speed	65mph

Production:

412 comprising 2 prototypes, 410 Mk I, Mk X (22 Mk I conversions)

Hamilcar carrying capacity

The variety of equipment capable of being carried by the Hamilcar was impressive. However, each variation in load required careful attention to the internal anchorage equipment. There could be no movement or shifting of heavy loads during flight. A military load of up to 17,500lb (or 7.8 tons) included:

1 x Tetrarch Mk IV tank
1 x Locust tank (US)
2 x Bren Gun Carriers
3 x Rota trailers
2 x Armoured scout cars
1 x 17-pounder anti-tank gun with portee vehicle
1 x Self-propelled Bofors gun
Jeep and Universal Carrier with slave batteries
1 x Universal Carrier for 3in mortars and 8 x motorcycles
Bailey pontoon bridge equipment
48 x panniers of equipment and ammunition

Airfield construction equipment:
D4 tractor with Angledozer
Scraper with Fordson tractor
1 x Grader
HD10 bulldozer (carried in 3 Hamilcars)
HD14 bulldozer (carried in 3 Hamilcars)

Waco CG-4A

Although the US Army used the British Horsa glider in limited numbers, the mainstay of US airborne operations was the Waco CG-4A ('Waco' standing for Weaver Aircraft Company of Ohio and 'CG' for cargo glider). Known in RAF service as the Hadrian, the CG-4A was built in the USA by 16 prime contractors (this number having risen to 23 by the war's end), with components being made by about 50 subcontractors, such as Steinway & Sons (the famous piano builders) who made wings and tail surfaces; H.J. Heinz (of '57 varieties' fame) who made wings and spar caps; and the Anheuser-Busch Brewery (better known as the brewers of Budweiser) who made inboard wing panels.

Some 13,903 examples of the CG-4A were built – more, for example, than the B-17 Flying Fortress – which made it the largest production run of any glider of the period.

It was constructed using a tubular steel frame with fabric covering, wooden wings with spruce main spars, and plywood internal flooring. Its simplicity meant that companies with no previous manufacturing experience could build the airframe – all they needed was the ability to cut and weld tubing and to work wood. Few manufacturers did all the work in-house, and most prime contractors took responsibility for the fuselages and steel work, while all the woodworking was subcontracted to those with expertise in wood, notably furniture manufacturers. The parts would be shipped in from the various subcontractors and were assembled at the main plant. Finished gliders could be taken apart and packed into five wooden crates for shipment overseas.

The need for woodworking capability is one reason why so many CG-4As were built in the American Great Lakes state of Michigan, where they were supported by the indigenous furniture industry. Between the Ford plant at Kingsford and Gibson's factory at Greenville they built 5,270 gliders.

For the purposes of loading and unloading, the complete nose section – including the pilots' seats – hinged up and out of the way so the cargo could be loaded straight in simply by pulling a couple of pins. A cable ran from the

LEFT The Waco CG-4A, the most widely used US glider of the war, was a study in simplicity. One commentator said of it: 'In concept, the fuselage is nothing more than a huge tubing box with the nose vaguely blunted and gigantic Hershy bar wings attached.' (USNA)

BELOW US glider-borne troops seated inside the tubular metal and fabric fuselage of a Waco CG-4A. (USNA)

ABOVE Early CG-4As suffered from frequent failures of the front landing gear fittings, the towline release mechanism and the nose-raising locking device, most of which were caused by faulty routine maintenance. Such problems should have been anticipated because the CG-4A had been designed and built for only one, non-return, flight into combat, and not for the hundreds of hours that many completed during training. (USNA)

top of the hinged nose section back along the top of the fuselage, turned through a pulley and was then attached to the back of the Jeep or howitzer inside the glider. As soon as the glider touched down, a latch was tripped on the nose section, so if the load broke free and tried to exit the front of the glider, its movement forward would yank the nose and the pilots up and out of the way. If the landing was a normal one, the Jeep would simply drive out the front and pull the nose up as it did so.

The CG-4A could carry 15 fully equipped troops, plus two crew. A normal load was 7,500lb, but in extreme combat conditions it could manage up to 9,000lb. It was designed for a maximum speed of 150mph, with a normal glide (descent) speed of 72mph and a landing speed of 60mph. The CG-4A's smaller size and lower landing speed than the Horsa meant it could land more safely and in tighter spaces (it could land and stop in 200 yards). This meant the CG-4A was better suited to touching down in the pocket-handkerchief size fields of Normandy, and its cargo of troops was less likely to sustain injury during the landing. It was also possible for a CG-4A waiting on the ground to be 'picked up' by an overflying C-47 using a simple net and tail-hook system. The usual tug aircraft were the Curtiss C-46 Commando and the C-47.

BELOW The nose of the CG-4A was fully hinging, complete with pilots' cockpit, giving access to the fuselage cargo bay. *(USNA)*

Specification –
WACO CG-4A (HADRIAN)

Crew	2
Capacity	17,500lb maximum payload: 15 fully equipped troops or an M3A1 75mm howitzer and crew of 3. British forces in the Mediterranean carried either a Jeep or a 6-pounder anti-tank gun and crew. Occasionally other vehicle loads were also carried.
Length	48ft 3¾in
Wingspan	83ft 8in
Height	15ft 4in
Empty weight	3,750lb
Max loaded weight	7,500lb

Performance:

Max towing speed	120mph
Stalling speed	50mph
Production	13,903

LEFT CG-4As on high tow behind C-47 Skytrains. *(USNA)*

BELOW The snatch launch recoveries of Waco and Horsa gliders relied in part on the elasticity of the nylon tow rope, along with a slipping clutch on the winch (think fishing reel) fitted to the Dakota towing aircraft to take the glider from stationary to flying in a few seconds. *(USNA)*

By Horsa to Normandy

Lieutenant-Colonel Iain Murray commanded 1 Wing, Glider Pilot Regiment. With co-pilot Lieutenant Bottomley, Murray flew to Normandy from RAF Harwell with the First Lift during the early hours of 6 June. His Horsa carried Brigadier Hugh Kindersley, with his personal Jeep and members of Headquarters 6th Airlanding Brigade, and the BBC War Correspondent Chester Wilmot, who provided a running commentary on the flight to a tape recorder. The excerpt that follows is taken from the transcript of this commentary, which Wilmot included in his masterly account of the last 11 months of the war against Germany, The Struggle for Europe. *When Murray landed his Horsa on LZ-N it was not without incident: one of the anti-glider obstacles on the landing field ripped off the left wingtip, while another pole collided head-on with the cockpit. Fortunately the pole was loose and was ripped from the ground on impact.*

Three o'clock: half an hour to go. The clouds clear for a minute and we are warned of the closeness of the coast – and of another tug and glider which has cut across our bow, perilously near. Away to our left the RAF is bombing enemy batteries near Le Havre and the sky is lit by the burst of bombs and flash of guns until the clouds shut us in again. Now, when we need a clear sky, it is thicker than ever and at times we lose sight even

of the tug's tail light. Suddenly the darkness is stabbed with streaks of light, red and yellow tracer from the flak guns on the coast. There are four sharp flashes between us and the tug and then another that seems to be inside the glider itself. It is, but we don't realise at first that we have been hit, for the shell has burst harmlessly well aft beyond the farthest seats. The tug begins to weave but it can't take violent evasive action lest the tow rope should snap.

Over the coast we run out of cloud, and there below us is the white curving strand of France and, mirrored in the dim moonlight, the twin ribbons of water we are looking for – the Orne and the Canal. The tug has taken

LEFT Heavily laden British paratroops inside an RAF DC-3 Dakota en route to their drop zone. (*IWM TR1662*)

us right to the target, but we can't pick out the lights which are to mark the landing zone. There is so much flak firing from the ground that it's hard to tell what the flashes are, and before the pilots can identify any landmarks we are into the cloud again.

Soon, one of them turns and calls back to us – 'I'm letting go, hold tight.' As it leaves the tug the glider seems to stall and to hover like a hawk about to strike. The roar of the wind on the wooden skin drops to a murmur with the loss of speed and there is a strange and sudden silence. We are floating in a sky of fathomless uncertainty – in suspense between peace and war. We are through the flak belt and gliding so smoothly that the fire and turmoil of battle seem to belong to another world.

We are jerked back to reality by a sharp, banking turn and we are diving steeply, plunging down into the darkness. As the ground rises up to meet us, the pilots catch a glimpse of the Pathfinders' lights and the white dusty road and the square Norman church-tower beside the landing-zone. The stick comes back and we pull out of the dive with sinking stomachs and bursting ears. The glider is skimming the ground now with plenty of speed on, and is about to land when out of the night another glider comes straight for us. We 'take off' again, lift sharply and it sweeps under our nose. The soil of France rushes past beneath us and we touch-down with a jolt on a ploughed field. It is rough and soft, but the glider careers on with grinding brakes and creaking timbers, mowing down 'Rommel's asparagus'

ABOVE Gliders of the 6th Airborne Division on Landing Zone 'N' at Ranville, where Chester Wilmot's glider came down. *(Author's collection)*

and snapping off five stout posts in its path. There is an ominous sound of splitting wood and rending fabric and we brace ourselves for the shock as the glider goes lurching and bumping until with a violent swerve to starboard it finally comes to rest scarred but intact, within a hundred yards of its intended landing place. It is 3:32am.

Chester Wilmot,
The Struggle for Europe (Collins, London, 1952)

ABOVE A Horsa glider near the Caen Canal bridge at Benouville on 8 June 1944. This was one of the gliders that carried men of 'D' Company, 2nd Oxfordshire and Buckinghamshire Light Infantry, to capture the bridges over the River Orne and Caen Canal on D-Day. *(IWM B5232)*

ABOVE Not all glider landings were successful. This is the smashed wreckage of an American Horsa, which almost certainly claimed the lives of some (if not all) of its passengers and crew. *(USNA)*

Glider tugs

Without a means of towing the gliders into battle they were going nowhere, so on 15 January 1942 the RAF formed 38 Wing in Army Cooperation Command to provide aircraft and crews for glider tug duties; but the RAF had no dedicated airborne forces aircraft in its inventory. From an inauspicious start in June that same year with a squadron of retired Whitley bombers, two years later on the eve of D-Day the RAF had made good with a selection of more hand-me-down bombers that had been converted from their original roles to carry paratroops and tow gliders – the Handley Page Halifax, Short Stirling, and the rejected bomber (turned paratroop-transport-cum-glider-tug), the Armstrong Whitworth Albemarle.

The Short Stirling and some earlier Merlin-engined versions of the Handley Page Halifax were heavy bombers that had been relegated to second-line duties owing to performance shortcomings and high loss rates. To equip them for their new roles, these aircraft were fitted with glider-towing apparatus and converted internally to carry and dispatch paratroops. Thanks to its capacious fuselage, the Stirling was used equally in both roles, whereas the Halifax was used primarily as a glider tug (it was the only tug aircraft capable of towing the giant Hamilcar glider).

In June 1942, the USAAF formed Troop Carrier Command, which was renamed the following month as 1st Troop Carrier Command (I TCC). Its primary role was to provide air transport for US parachute, glider and airborne infantry troops with all their equipment and supplies. The Douglas C-47 Skytrain was the principal paratroop and glider tug aircraft, although the Curtiss C-46 Commando was also used, but in smaller numbers.

Short Stirling

When the Air Ministry realised that the Stirling lacked any potential for future development as a bomber, it was considered as a possible glider tug and paratroop transport to replace the Albemarle and Whitley. Large numbers of the Stirling Mk III were converted on the factory production lines, or at RAF Maintenance Units, into the Mk IV for use in these roles. The changeover to completed Stirling Mk IVs began in December 1943.

The new Mk IV differed from previous builds of Stirling in many ways, the most obvious being the removal of the nose and dorsal power-operated gun turrets, and the replacement of the nose turret with a transparent Perspex fairing. A 6ft x 4ft hatch in the fuselage floor just forward of the existing crew entrance door was provided for paratroop dropping. A retractable tubular-metal-frame strop guard (or 'gate') was mounted beneath the rear fuselage behind the paratroop hatch, which was lowered beneath the fuselage and held in place by a strut and pin. The gate was used to trap the parachute static lines and prevent them from whipping up into the tailplane in the slipstream and causing damage. Once all paratroops had jumped, the static lines could be winched back on board by

BELOW RAF Keevil in Wiltshire, May 1944. No 196 Squadron's Stirling Mk IV, W-Whisky, tows off a Horsa. *(Author's collection)*

the dispatcher in the Stirling aircrew, and the guard retracted manually from inside by use of a raising/lowering arm in the fuselage floor.

For glider towing, a horseshoe-shaped metal towing bridle was fitted behind and below the horizontal stabilisers, with three points for connecting tow ropes and a provision to link into the tug's intercom by means of an audio lead and plug. An intercom wire was woven into the tow rope, connecting the glider and tug, but in practice this wire often broke during training through the stresses placed on it by towing, casting off and recovery of the rope for reuse.

Handley Page Halifax

When the Central Landing School (later Establishment) was opened at Ringway on 19 June 1940 to train paratroops and glider-borne troops, the aircraft spared by the RAF from its main task for a secondary role as paratroop transport and glider tug was the Armstrong Whitworth Whitley bomber. This role fell to 4 Group Bomber Command, and when its Whitleys were replaced in 1941 by Halifaxes, the latter were brought into the airborne role.

The first Halifaxes allocated to the Central Landing Establishment arrived in October 1941 and were modified for paratroop dropping. Other Halifaxes were involved in glider towing. Although the Whitley was capable of a short-haul tow with a Horsa, a Halifax or similar four-engined aircraft was vital for any long haul. The Short Stirling's large box-like fuselage

ABOVE The horseshoe-shaped glider-towing bridle fitted to the rear fuselage of a Stirling Mk IV. Note the intercom wire attached to the bridle. (Author's collection)

LEFT The Halifax was the only aircraft capable of towing the giant Hamilcar glider. These are Halifax Mk Vs of 644 Squadron lined up at Tarrant Rushton in June 1944. (T.A. Pearce/Andy Thomas)

RIGHT The crowded runway at Tarrant Rushton on 5 June, packed with Hamilcar gliders and their Halifax tugs of 298 and 644 Squadrons. Heading the line-up is a pair of Horsas. *(Andy Thomas collection)*

BELOW Halifaxes and Stirlings (in the background) tow Horsas. *(Author's collection)*

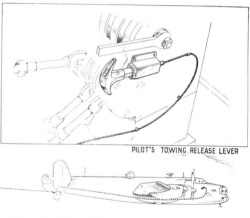

PILOT'S TOWING RELEASE LEVER

TOWING AND RELEASE UNIT (TYPE 6A)

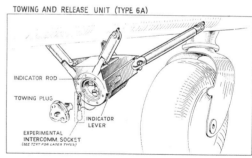

INDICATOR ROD

TOWING PLUG

INDICATOR LEVER

EXPERIMENTAL INTERCOMM. SOCKET
(SEE TEXT FOR LATER TYPES)

ABOVE Type 6A glider towing and release apparatus as used on the Halifax Mk II and V. *(Author's collection)*

lent itself more conveniently to the carriage of paratroops, leaving the Halifax as the obvious choice for glider towing. When the giant Hamilcar glider came into service only a four-engined aircraft like the Halifax could handle its great weight.

For the D-Day operation, 38 Group was required to provide four Stirling, four Albemarle and two Halifax squadrons. The Halifax component was met by 298 and 640 Squadrons with 18 plus 2 tug aircraft, and 70 Hamilcar and 50 Horsa respectively glider establishment. An extra complication was that when towing the Hamilcar, the Rolls-Royce Merlin 20 engines of the Halifax Mk II and V had to be replaced with Merlin 22s.

Douglas DC-3 Dakota

The workhorse aircraft of the US 9th Troop Carrier Command (IX TCC) was the Douglas C-47 Skytrain (DC-3 Dakota in RAF service), the military version of the highly successful and dependable DC-3 civilian passenger airliner. On and after D-Day the IX TCC used the Skytrain almost exclusively for its airborne forces operations: 50th Troop Carrier Wing (439th, 440th, 441st, 442nd Troop Carrier Group); 52nd Troop Carrier Wing (61st, 313th, 314th, 315th, 316th, 434th Troop Carrier Group); 53rd Troop Carrier Wing (61st, 435th, 436th, 437th, 438th, 439th, 440th Troop Carrier Group).

The RAF also used the Dakota during the Normandy campaign (and later) as a paratroop transport and for glider towing. No 46 Group commanded five RAF squadrons of Dakota Mk IIIs dispersed over three airfields: Down Ampney, Gloucestershire (48 and 271 Squadrons); Broadwell, Oxfordshire (512 and 575 Squadrons); Blakehill Farm, Wiltshire (233 Squadron).

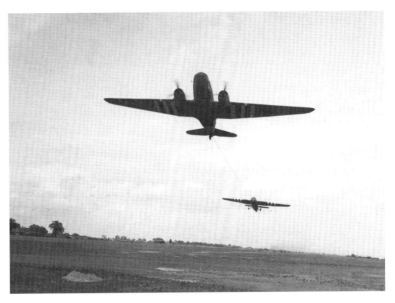

TOP A C-47 of the USAAF's 90th Troop Carrier Squadron, 438th Troop Carrier Group, tows off Horsa '32' from Greenham Common. *(USNA)*

ABOVE Dakota Mk IIIs of the RAF's 46 Group at B2/Bazenville, loading casualties for evacuation to the UK. Identifiable aircraft include KG432 'H' of 512 Squadron (centre), and KG320 'B' of 575 Squadron (extreme right). *(IWM CL3885)*

LEFT The American C-47 could carry 27 fully equipped paratroopers seated inside the fuselage. *(Author's collection)*

RIGHT The Armstrong Whitworth Albemarle was not loved by paratroopers because of its cramped interior. V1823, P5-S, of 297 Squadron (pictured) was piloted on the night of 5–6 June by Flying Officer Edward Halpin, towing Horsa glider '35', flown by Staff Sgt Colin Hopgood, carrying a 13th Parachute Battalion Jeep and trailer, motorcycle and five soldiers. The Horsa was destined for Ranville but its tow rope broke en route, causing it to crash-land at Saint-Vaast-en-Auge in the early hours of 6 June. The glider crashed into trees, killing Staff Sgt Hopgood and his co-pilot, Sgt Daniel Philips, and three of the soldiers. (Author's collection)

Armstrong Whitworth Albemarle

The Albemarle was a unique but lacklustre aircraft, conceived as a bomber but used as a general transport, built of wood and steel instead of the usual light alloys to meet anticipated shortages of light alloys and specialist aircraft manufacturing facilities.

The Albemarle ST Mk I Series 2 was equipped with Malcolm Glider Towing gear; the Mk II and Mk V could carry 10 fully equipped paratroops or tow a glider. There was a dropping hole in the floor of the rear fuselage and a large loading door in the fuselage side.

The tug and glider tow-points were manufactured by R. Malcolm & Co. Ltd of Slough, whose other contributions to the Allied war effort included the bulged Malcolm Hood cockpit canopy on the P-51B and C Mustang, and the F4U Corsair, and fuselage and tail-feathers for the Haffner Rotabuggy flying Jeep.

On D-Day, four RAF squadrons in 46 Group operated the Albemarle – 295, 296, 297 and 570. A handful of other squadrons also operated small numbers of the aircraft.

The Albemarle's fuselage interior was cramped, making it unpopular with paratroopers, and its two Bristol Hercules XI radial engines were prone to overheating when towing a glider.

RIGHT Albemarle paratroop installations looking aft. (Author's collection)

FAR RIGHT Albemarle paratroop modifications. (Author's collection)

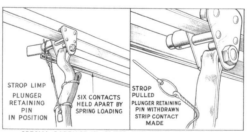

STROP LIMP
PLUNGER RETAINING PIN IN POSITION

SIX CONTACTS HELD APART BY SPRING LOADING

STROP PULLED
PLUNGER RETAINING PIN WITHDRAWN STRIP CONTACT MADE

SPECIAL CARRIAGE (No.5) FOR OPERATING STROP SWITCH

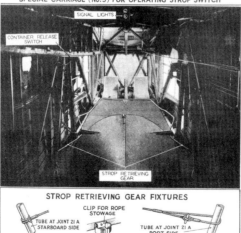

SIGNAL LIGHTS

CONTAINER RELEASE SWITCH

STROP RETRIEVING GEAR

STROP RETRIEVING GEAR FIXTURES

CLIP FOR ROPE STOWAGE

TUBE AT JOINT 21 A STARBOARD SIDE

TUBE AT JOINT 21 A PORT SIDE

FIG.15 **ALBEMARLE** PARATROOP INSTALLATIONS LOOKING AFT

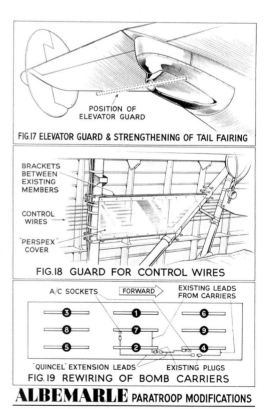

POSITION OF ELEVATOR GUARD

FIG.17 ELEVATOR GUARD & STRENGTHENING OF TAIL FAIRING

BRACKETS BETWEEN EXISTING MEMBERS

CONTROL WIRES

'PERSPEX' COVER

FIG.18 GUARD FOR CONTROL WIRES

A/C SOCKETS   FORWARD   EXISTING LEADS FROM CARRIERS

③ ① ⑥
⑧ ⑦ ⑨
⑤ ② ④

'QUINCEL' EXTENSION LEADS   EXISTING PLUGS

FIG.19 REWIRING OF BOMB CARRIERS

ALBEMARLE PARATROOP MODIFICATIONS

Just a piece of rope
Leading Aircraftman (LAC) Alf Sore,
1 Heavy Glider Servicing Unit, Netheravon, 1944

When LAC Alf Sore heard he was being posted from a cold RAF Thornaby-on-Tees, with its troublesome Vickers Warwicks, to RAF Netheravon on Salisbury Plain, he was at first very pleased, but on learning he was 'detailed' to the Rope Section his heart sank.

The Rope Section comprised a small workshop in the corner of a hangar filled with hundreds of ropes all neatly stacked, each with different labels attached, but as I got interested I found it was much better than I anticipated.

There were two types of rope, the smaller was 350ft long by 3½in circumference (rope is measured in 'circ' not 'dia') and these were used to tug Horsa and Waco/Hadrian gliders. The larger ropes, still 350ft long but 4½in circumference, were used to tug our biggest glider, the Hamilcar. Each rope had its own log book recording its serviceability and number of tugs, a label with the rope's number and, on a new rope, ten pieces of tape threaded between its strands. After each tug one tape was removed, the rope inspected, splices checked and details entered in its log book. All ropes had a radio cable threaded through to allow the tug and glider pilots to communicate.

The greatest amount of work we had to do was on the connecting fittings spliced to each end of the rope. The English and American couplings were different (what's new?). The American one was basically a big hook and eye, but the English had what was called 'Lobel' fittings. The easiest way I can describe it is to clench one hand into a fist and then clasp it with the other hand. To release, was of course to open the second hand. Both tug and glider were able to make the release but, other than in an emergency, it was always the glider.

It was a constant job changing these fittings. We (the English) had four types of tug: Whitley, Halifax, Stirling and Albemarle, which all had standard 'E'-type fittings. The Hadrian had 'A'-type fittings and the Horsa type 'E'. Of course the American tugs differed. For instance, the Dakota had 'A'-type to the Hadrian's 'A', and the Horsa's 'E'. There seemed to be endless combinations but the easiest pair to connect was

ABOVE Leading Aircraftman Alf Sore, No 1 Heavy Glider Servicing Unit. *(Alf Sore)*

the Halifax (the most powerful tug) to the Hamilcar, because both fittings were the same.

Another change came when the Horsa Mk II came into service, which had a 'straight' rope pull from the tail of the tug to the nose of the glider. (The Horsa Mk I had a 'Y' pull from the tail of the tug to the glider's main planes, or wings.)

Of course questions were always asked following rope breaks, but the answer was that they were never due to a lack of maintenance. They usually occurred when the tug and glider were not in line, mostly caused when the glider pilot lost sight of the tug in cloud. The glider had an instrument in the cockpit to show the angle of the tug to the glider (called the 'angle of dangle'), but this was not very good if the angles changed very quickly.

Another reason for rope breaks was when the glider changed from 'high tug' to 'low tug'. When the glider flew through the tug's slipstream a 'push-pull' effect occurred and the 'chucking' by this would cause the rope to break.

Radio homing devices – Rebecca and Eureka

During the Second World War, accurate insertion of airborne forces was always difficult, but in 1944 TRE developed the Rebecca and Eureka system of airborne direction finding equipment and portable ground-based beacons to assist in air-dropping supplies to the Allied armies and to Resistance groups in Occupied Europe. Rebecca was the airborne station, and Eureka the ground-based beacon.

The Eureka beacon had a 5ft retractable mast on its 7ft tripod, which was connected to a sealed box that only needed to be switched on for the signal to transmit. Its signal could be activated only when the incoming aircraft transmitted a coded signal from its Rebecca aerial.

The beacon signal might transmit for ten minutes if it was activated at the maximum range of 60km. Accuracy of the Eureka signal was such that it allowed aircraft navigators to pinpoint its position to within 200m, giving them a bearing to and distance from the drop zone (DZ, for paratroops) or landing zone (LZ, for glider-borne troops and equipment).

Preliminary trials with the Eureka II version had shown that it was too heavy to be used by Parachute Regiment pathfinders, so a lighter-weight version (Eureka III) was designed and developed in conjunction with A.C. Cossor using American 9000-Series miniature valves. The British Eureka III, including 6-volt batteries, was carried in a webbed harness around the waist of a paratrooper who had only to unclasp the harness, remove and erect the telescopic aerial, to operate the equipment.

US paratroops used the American-developed AN/PPN-1, AN/PPN-2 (Portable) Eureka, which was based largely on the British Eureka III design.

RAF and USAAF paratroop transport aircraft used AN/APN-2 Rebecca, which was developed from the SCR 729 Airborne Interrogator, of which 1,000 sets were delivered before mid-1944 for use in the D-Day airborne operations.

For airborne forces operations, paratroopers known as pathfinders (the first paratroopers to land) were required to set up Eureka transmitter beacons on the DZ or LZ, in advance of the arrival of the main 'lift' (airborne force). Some pathfinders were assigned to mark the parachute DZs, while others marked the glider landing fields.

Once the pathfinders had set up the Eureka beacons for the troop carriers to beam in on, the team responsible for the lighting markers

RIGHT The H-shaped Rebecca aerial can be seen on the side of the nose of this 620 Squadron Stirling Mk IV. *(Author's collection)*

checked the wind direction and lined up the seven-light 'T' with the aid of an Automatic Direction Finder (ADF). The 'T' arrangement used seven Holophane prismatic lights placed 25yd apart – four lights were usually across the top and at least three more formed the stem. The direction of the jump was indicated by the stem of the 'T', while the crossbar of the 'T' marked the 'go' or 'jump' point.

The lights were clearly visible from the air, but almost completely hidden from the ground. The tail of the 'T' was programmed to wink out the signal of the DZ (eg drop zone 'A' blinked out the Morse code signal for the letter 'A'). The T-lights were also colour-coded according to the drop zone they were marking: DZ 'A' was amber, 'C' was green, 'D' was red. The Eureka beacon was located within a 100yd radius from the head of the 'T'.

On the eastern flank of the invasion area gliders of the British coup de main force landed beside the Caen Canal (LZ X) and the River Orne (LZ Y) at around 12:15am on 6 June, where they successfully took 'Pegasus' and 'Horsa' bridges. Shortly afterwards the pathfinders of the 22 Independent Parachute Company dropped from six Albemarles over DZs V, N and K to mark out the drop zones for the advance parties of paratroops soon to follow.

The advance party from the 8th Parachute Battalion (8 Para) landed successfully at DZ K, but at DZ V, west of Varaville, all of the Eureka beacons were smashed on landing. With no means of pinpointing the DZ, the advance parties from 3rd Parachute Brigade HQ, 9 Para and 1 Canadian Para were badly scattered. Six men jumped from one aircraft while it was still over the coast and were lost without trace; two more aircraft could not find the DZ and were hit by flak. This meant that when the main lift of 9 Para came in at 12:50am only 17 out of the 71 aircraft were able to drop their loads accurately.

The coup de main force destined for the assault on the Merville Battery comprised three Horsa gliders of B Squadron GPR fitted with Rebecca Mk III, towed by Albemarles of 297 Squadron, carrying 56 men and eight sappers of G-B Force. However, they were unable to pick up a Eureka signal; two gliders landed some distance from their LZ, while the third returned to England owing to technical troubles.

Rebecca at work

The airborne Rebecca equipment radiates five microseconds duration interrogating pulses on a VHF spot frequency. On receipt of the interrogating pulses, the Eureka ground-based beacon triggers its associated transmitter, causing responses to be radiated on a different frequency, but at the same Pulse Repetition Frequency (PRF) as the interrogating transmitter. The returned signals received in the aircraft by both the right and left Rebecca receiver aerials are displayed on a cathode ray tube (CRT) indicator unit.

If the beacon is to the right of the aircraft, the signal to the right of the time base will have the greater amplitude, in which case a right turn will be necessary to make the signals on either side of the time base equal in amplitude. This would indicate that the aircraft was then flying directly towards the beacon.

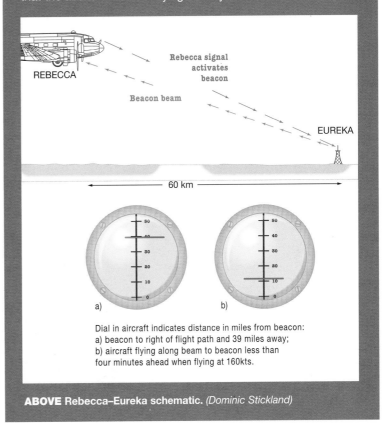

Dial in aircraft indicates distance in miles from beacon:
a) beacon to right of flight path and 39 miles away;
b) aircraft flying along beam to beacon less than four minutes ahead when flying at 160kts.

ABOVE Rebecca–Eureka schematic. (Dominic Stickland)

Thanks mainly to the accurate positioning of the Eureka beacons at DZ N, the most successful British paratroop drop on D-Day was east of the River Orne between Ranville and Breville, by 110 Stirlings and Albemarles and 21 Dakotas. Some 2,026 men of 5th Parachute Brigade and 702 ammunition and equipment containers were accurately dropped, with additional guidance from Holophane T-lights on the DZ and the moon glinting on the waters of the Caen Canal and the River Orne.

The effectiveness of some US pathfinder teams on the western flank around Sainte-Mère-Église and Carentan on D-Day was hindered when the incoming C-47 formations were scattered by low cloud and German flak. This caused many serials to be dropped away from their intended DZs, but overall the US airborne drop was a success. Even so, some of the US parachute battalions were so badly scattered that they took days to re-form.

By Skytrain to DZ O
Lieutenant-Colonel Edward Krause

Commanding the 3rd Battalion, 505th Parachute Infantry Regiment (3/505th PIR), US 82nd Airborne Division, 'Cannonball' Krause flew to Normandy from RAF Cottesmore in one of 36 C-47 Skytrains of the USAAF's 316th Troop Carrier Group (TCG). Their time over the DZ was planned as 1:57am. He describes the experience of parachuting on to drop zone O, south-west of Neuville-au-Plain, in the early hours of D-Day:

I would say that in the next three minutes I came as close to being crashed in the air as I ever hope to be. The pilot called for evasive action and we split up. Some went high, some went lower, others right and left. This split our formation and we were well spread. Just about two or three minutes before the drop time, we saw the green T [T-lights], it was a Godsend and I felt that I had found the Holy Grail. I would say that I dropped from over 2,000ft. It was the longest ride I have ever had in over 50 jumps, and while descending, four ships [aircraft] passed under me and I really sweated that out.
(Official record of an Operation Neptune debriefing conference, 13 August 1944, quoted in Battle Zone Normandy: Utah Beach.*)*

Cloud over the DZ meant that some C-47 pilots decided to drop from a greater height than usual. The pathfinders had made such a good job of marking the DZ that many pilots circled back over the area to drop their paratroopers more accurately. In fact, DZ O had the most precise series of jumps of any that night.

Once on the ground, Krause, with 180 men, struck out in the darkness for Sainte-Mère-Église. Using only knives, bayonets and hand grenades (so that any gunfire could be treated as hostile) they captured the town by 4:30am and succeeded in cutting the Germans' main communications cable with Cherbourg.

ABOVE At 11:10pm on 5 June, Albemarle V1740, 8Z-A, of 295 Squadron was the first aircraft to take off for Normandy carrying pathfinder troops of 22 Independent Parachute Company. In this photograph the troops prepare to board the aircraft; Lt Bob Midwood, one of the four stick commanders, is seen standing in front of the Z on the side of the aircraft, facing his stick of paratroopers. The Albemarle's pilot was Sqn Ldr Claude Merrick (standing in front of the '8'), with the AOC of 38 Group, Air Vice-Marshal Hollinghurst, as a passenger. Merrick dropped the paratroops from 500ft over DZ K at 12:17am. For their part in this operation, Merrick and his navigator Warrant Officer Robert Farrow (standing third from left) were each awarded the DFC. *(IWM H39071)*

RIGHT Lt-Col Robert L. Wolverton, commanding officer of the 3rd Battalion, 506th Parachute Infantry Regiment (PIR), 101st Airborne Division, and his Headquarters team, check their equipment before boarding C-47, 8Y-S, *Stoy Hora*, of the 98th Troop Carrier Squadron, 440th Troop Carrier Group, at Exeter on the evening of 5 June. *Stoy Hora* was the lead ship, chalk No 1, of the 98th TCS, piloted by Col Frank X. Krebs, the 440th TCG's commanding officer. Wolverton, 29, was killed by German machine-gun fire in an orchard outside Saint-Côme-du-Mont on 6 June. *(USNA)*

ABOVE Pathfinders of 22 Independent Parachute Company, 6th Airborne Division, blacken their faces in front of an Albemarle aircraft at RAF Harwell on 5 June 1944. *(IWM H39066)*

LEFT US paratroopers of the 101st Airborne Division check each other's kit before take-off. *(USNA)*

RIGHT US paratrooper pathfinder badge. *(Author)*

BELOW On the evening of 5 June, paratroopers of the 508th PIR check each other's kit on the airfield at Saltby, Leicestershire, before being flown to DZ N by C-47s of the 314th Troop Carrier Group. (USNA)

ABOVE Men of F Company, 2nd Battalion, 506th PIR, wait for take-off in C-47, chalk No 12, of the 439th TCG from Upottery, Devon. At 1:20am they jumped over DZ C (Hiesville). From left to right: William G. Olanie, Frank D. Griffin, Robert J. Noody (holding a bazooka), and Lester T. Hegland. When Bob Noody jumped from the C-47 he weighed about 250lb (almost 18 stones) with his M-1 rifle, a bazooka, three rockets, land mines and other assorted kit. From his chest hangs 59ft of bundled rope, used (unravelled) to lower his leg bag to the ground, which eased his fall and ensured he was ready to fight when he landed at Sainte-Mère-Église. (USNA)

ABOVE Remembering the events of 1944, military parachutists drop over Normandy on the 60th anniversary of D-Day. (US Dept of Defense)

HEADQUARTERS 82D AIRBORNE DIVISION
Advance Command Post
APO 469 – In the Field
11 June 1944

Subject: Report of Pathfinder Employment for Operation 'NEPTUNE'
To: Commanding General, 82D Airborne Division

1. Pathfinder teams of the 505th, 507th and 508th Parachute Infantries were scheduled to be employed on their respective DZ's at H-30, D-Day, in accordance with Field Orders of these organisations.
   The Regimental Pathfinder Teams consisted of three battalion teams composed of two officers, two Eureka operators, one wire man, seven light men, and from four to six security men. The 507th and 508th Pathfinder teams had four security men per each battalion assigned from the 504th Parachute Infantry, plus one officer for each Regimental Team.
   The 508th Regimental Team, in addition to the above equipment and personnel, dropped two BUPS beacons [AN/UPN-1 Beacon Ultra Portable S-band – radio direction finding beacon, not as effective as Rebecca-Eureka] plus the Commanding Officer of the Provisional Pathfinder Company.

2. All regimental Pathfinder teams flew in 3-plane flights, in V-formation. Take-off aerodrome was NORTH WITHAM.

3. The 505th Pathfinder team, commanded by 1st Lt JAMES J. SMITH, Air Corps Commanding Officer Captain KIRKPATRICK, took off as per schedule.
   Encountering little flak on the run-in from landfall to DZ, dropped within 400 yards of previously designated pinpoint. The drop was six minutes early. The DZ was set up as per SOP [standard operating procedure], with the exception of one battalion light tee. This tee was not put into action due to the faulty assembly. Eurekas were set up within ten minutes of drop time. Eurekas were first triggered fifteen minutes from the time of the first element drop. Three serials dropped on and near the 505th pinpoint. All serials were approximately ten minutes early. The first serial dropped southeast of lighted tee, approximately one-half mile. The second serial dropped directly over lighted tee. The third serial dropped north by approximately one-half mile.
   The third serial appeared to be travelling at a speed of at least 150 miles per hour, at time of drop. Several jumpers sustained ruptures, due to the excessive speed.

4. LZ's in 505 area were set up as per schedule, Eurekas for glider night landings were set up thirty minutes prior to landing time. Eurekas for LZ were triggered twenty minutes prior to landing time. Gliders were generously scattered over LZ area.

5. The 507th, commanded by 1st Lt. JOSEPH, 507th Parachute Infantry, flight led by Captain MINOR, Air Corps. Flight took off on time at NORTH WITHAM. Dropped on designated DZ accurately, on time. At time of drop, pathfinder personnel and pathfinder aircraft were subjected to heavy anti-aircraft fire. The jumpers, on reaching the ground, found themselves in a German troop concentration. Due to aggressive action of enemy troops, the DZ was not set up according to SOP. No lights were turned on. One Eureka was set up by this Pathfinder team twenty minutes prior to scheduled drop of first serial. Eurekas were triggered fifteen minutes prior to drop of first scheduled serial. All elements appeared on time. These elements were widely scattered upon arrival, apparently due to action of enemy anti-aircraft fire. A maximum of fifty aircraft dropped their parachutists on the DZ. Eureka remained in action twenty minutes after scheduled time of last serials. A few strays dropped after scheduled time.

6. 508th, commanded by Captain N.L. MC ROBERTS of the 505th Parachute Infantry, Air Corps flight leader, Captain MILES, took off on time, from NORTH WITHAM, made landfall on time, encountering little flak until over SAINT-SAUVEUR-LE-VICOMTE. Flak continued from SAINT-SAUVEUR-LE-VICOMTE to run-in for drop. Anti-aircraft fire shifted from planes to jumpers at time of drop. Drop was on time, approximately one and one-half miles south and slightly east from previously selected DZ. Due to aggressive enemy action on the ground, lights were not able to be turned on with the exception of two; one of which was coded in the pre-designated code. BUPS Beacon was set up and operating twenty minutes prior to arrival of first scheduled serial. One Eureka was set up and operating twenty minutes prior to first scheduled serial. Eureka was triggered approximately twelve minutes prior to drop time. BUPS Beacon was receiving definite tuning of homing planes. To ground observers it appeared that incoming formations were scattered due to intense anti-aircraft fire. One large formation was observed dropping approximately one mile directly north. Twenty planes dropped on DZ with pathfinders. Twenty planes that dropped were approximately ten minutes late, of the first scheduled serial. No subsequent serials arrived over DZ. Eureka remained on thirty minutes after time of last scheduled serial. No strays dropped during that time.

7. 505th Parachute Infantry sustained no casualties due to enemy action. 508th Pathfinders lost approximately two-thirds of their enlisted and officer personnel. 507th Parachute Infantry is missing approximately twenty men. All navigational aids used by Pathfinders were recovered and consolidated in Division CP with the exception of those that were destroyed to avoid their falling into enemy hands.

8. For future Pathfinders operations it is recommended:

   a. That lights either be entirely eliminated or of such construction that they are not visible from the ground.
   b. That Pathfinder teams stress in their training assembly under difficult terrain conditions at night.
   c. That the size of Pathfinders teams be considerably lessened.
   d. That security personnel be dispensed with.
   e. That Air Corps pilots and crews be trained to such a degree that formations will be maintained in spite of intensive enemy anti-aircraft fire.
   f. That the BUPS beacon, both antenna and receiver-transmitter set, be modified in a more compact unit for jumping.

M.L. MC ROBERTS, Captain, Infantry, 82d A/B Div Pathfinders.

ASSAULT FROM THE SKIES

(Source: http://www.6juin1944.com)

Chapter Six

A warrior reborn

The Assault Glider Trust's Horsa

Inside a hangar at an RAF base in the north-west of England, a team of volunteers has painstakingly scratch-built a Horsa glider using original plans and by examining surviving aircraft components. The aircraft has been constructed using original techniques, and parts from salvaged Horsas have been incorporated wherever possible.

OPPOSITE The Assault Glider Trust Airspeed Horsa Mk I has been scratch-built by a team of volunteers working from a hangar at RAF Shawbury. *(All photographs © Assault Glider Trust)*

British military gliders in the Second World War were named after historical military figures whose names begin with the letter 'H' – Hengist, Horsa, Hamilcar and Hadrian. The brothers Hengist and Horsa were two such figures of Anglo-Saxon legend who led the Angle, Saxon and Jutish armies in their conquest of the first territories in Britain during the 5th century. These warriors lent their names to the Slingsby Hengist and the Airspeed Horsa gliders.

Formed out of the Assault Glider Association, the Assault Glider Trust (AGT) is a registered charity whose origins go back to the year 2000, when veterans of the Midland Branch of the Glider Pilot Regiment Association decided to construct a complete Airspeed Horsa assault glider to serve as a fitting memorial to airborne forces in the Midlands.

Of the several hundred Horsas used in the Second World War, many were built in Birmingham and were assembled and tested at West Midland airfields like RAF Cosford, Shawbury and Sleap, before delivery to the Glider Pilot Regiment's operational units.

Surprisingly, no complete example of the Horsa has been preserved anywhere in the world. While honouring two Midland glider-borne regiments, the South Staffords and the Ox & Bucks Light Infantry, both of which acquitted themselves with outstanding gallantry, no memorial would be complete without reference to the RAF crews who towed the gliders, dropped paratroops and kept ground forces resupplied during the battle for Europe.

The great airborne battles in north-west Europe and the Mediterranean are well documented, but the Burma campaign should not be forgotten. Six brigades of Chindits were inserted behind Japanese lines, many by glider, and were resupplied almost entirely from the air. The smaller, metal-framed American Waco CG-4A glider was used in these operations.

In June 2001, RAF Station Shawbury in Shropshire offered hangar space for the construction of an Airspeed Horsa Mk I by a team of volunteers who lived in the local area; the station continues to support the volunteer team. Initially a section of Horsa fuselage, recovered after the war from a landing site in Normandy, was loaned for the volunteers to copy. One year later, some copies of the original working drawings were donated by BAE Systems to allow the Horsa construction to proceed, provided an assurance was given that the new aircraft would not be flown.

In 2004, the Trust acquired a Douglas DC-3 Dakota aircraft as a gift from Air Atlantique, Coventry, and funds were raised to convert it back to its former glider-towing and para-dropping role, and to give it back its wartime identity.

In 2005, the Trust received a consignment of CG-4A Waco parts from the Silent Wings Museum at Lubbock, Texas. The fuselage and cockpit have now been completely reassembled by a small team of volunteers working alongside the Horsa. The Trust completed its collection of aircraft in 2007 with the acquisition of a DH-82 Tiger Moth, used to train glider pilots in the 1940s, which has now been fully restored.

In 2012, the main fuselage of the Horsa was virtually complete, with future work concentrating on the outer wing sections. Similarly, the Waco team were focusing their efforts on wing construction. Work continues on the Dakota's internal fittings, as does refurbishment of the Tiger Moth's engine.

Ultimately the Trust aims to put its aircraft and associated weapons, vehicles, uniforms and other memorabilia on permanent display for the general public, ideally as part of an existing museum in the Midlands, to serve as a permanent memorial to all those Service personnel, both Army and RAF, and civilians involved in assault glider operations in the Second World War.

The Assault Glider Trust's Airspeed Horsa has been built from scratch using original plans and by examining surviving aircraft components. Although the plans show a great deal of detail, they are mostly 'assembly' drawings and are often lacking measurements and sizes. This has required the volunteer workforce to do a lot of 'reverse engineering' in order to make everything fit together correctly. The aircraft has been built using original techniques, and original parts have been incorporated wherever possible. Although the Horsa is being built to flight condition, it will not be flown.

1   Measurements were taken from an original Horsa fuselage to assist with drawing up accurate plans for the AGT's new-build Horsa.

2   Assembly plans from the original Airspeed Horsa Air Publication were helpful, but without the all-important measurement details they were more of a guide than a comprehensive set of instructions.

3   The circular fuselage bulkheads are fabricated.

4   The fuselage takes shape.

5   A view towards the rear fuselage. Joints between the barrel sections were lapped towards the tail, which involved expanding the aft end of each section to fit over the one immediately behind it.

6   Assault Glider Trust volunteers at work in the hangar workshop at RAF Shawbury.

7   The rudder in its simple jig.

8 The rudder post and rudder have been attached to the fuselage.

10 WISA-Craft GL I birch plywood skins (1.5mm thick) are bonded to the solid wood frames with Aerolite 300 synthetic urea-formaldehyde adhesive. Nylon straps hold it tightly in place until the adhesive has fully cured and the ply is firmly bonded to the structure. This specialist thin veneer plywood is well suited to the technically challenging task of fuselage cladding.

9 A volunteer is dwarfed by the huge rudder structure, which is seen here before the attachment of its linen fabric skin.

11 This is the aperture in the forward port side of the fuselage where the downward-hinging main cargo door will be fitted.

12 The skin panel joins have been sealed. Note the main undercarriage units that are now attached, and the wing main spar attachment bracket on the upper fuselage.

13 A Night Black topcoat is then applied and the fuselage roundel planned out, for which the painting has started. The horizontal stabilisers have still to be fitted.

14 Building the cockpit section.

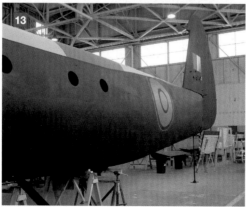

15 Fabricating the
   pilot's control
   column.

16 The cockpit section
   before the glazing
   has been fitted.

Airspeed Horsa instrument panel and controls – key

1 Pilot and co-pilot control columns
2 Wing nuts securing control wheels
3 Control wheels
4 Rudder bars
5 Altimeter
6 Air brake control levers
7 Airspeed indicator
8 Air pressure gauge
9 Flying limitation plate

10 Instrument panel light
11 Artificial horizon
12 Rate of climb and descent indicator
13 Turn and bank indicator
14 Elevator tab control
15 Flaps control lever
16 Tow release control lever
17 Undercarriage jettison control lever
18 Compass

17 Building the cockpit section.

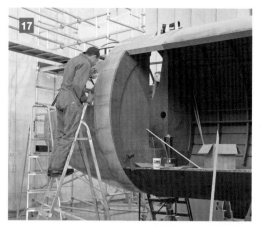

18 The cockpit section has its glazing fitted.

19 A scaffolding platform is put in place to fit the wing centre section and undercarriage struts.

20 LH291 stands on its landing gear while the inner wing sections are attached.

RIGHT General-assembly instructions for the Horsa.

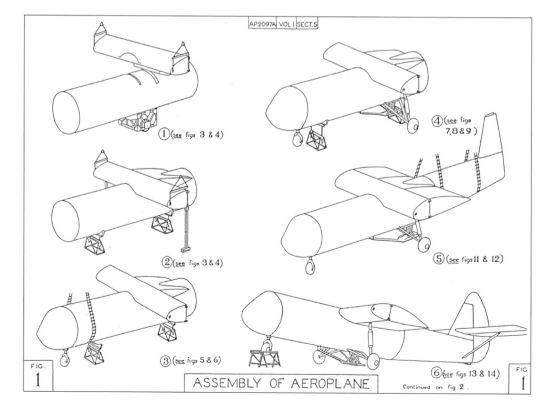

LEFT Visitors to RAF Shawbury inspect the finished fuselage of the glider.

BELOW Flat-pack Horsa. Civilian workers pose with sections of a Horsa glider, as received from the manufacturers, before it is assembled at 6 Maintenance Unit, RAF Brize Norton in Oxfordshire, during April 1944. *(IWM CH13022)*

Chapter Seven

Fields for tactical air power

Building Advanced Landing Grounds

Dozens of temporary airfields were built by RAF and USAAF combat engineer teams across Normandy after D-Day for use by Allied tactical air forces in support of the advancing armies. Within 24 hours of the landings three such airfields had been set up and were operational. The rapid construction of these Advanced Landing Grounds was a triumph of civil engineering.

OPPOSITE Ground crew watch Typhoon Mk IBs of 175 Squadron as they taxi out for a sortie at B5/Le Fresne-Camilly, Normandy, on 7 July 1944. *(IWM CL403)*

ABOVE Framed by the entrance of a Butler combat hangar draped with camouflage netting, Spitfire LF Mk IXBs of 312 and 313 (Czech) Squadrons undergo engine repair and maintenance at Appledram Advanced Landing Ground (ALG) in Sussex, on 19 April 1944. *(IWM CH18720)*

RIGHT The Commander-in-Chief of the Allied Expeditionary Air Forces, Air Chief Marshal Sir Trafford Leigh-Mallory, chats with Flt Sgt Joe Fargher at B2/ Bazenville ALG in Normandy. Fargher was one of three 234 Squadron Spitfire pilots forced down in the Caen area on 14 June after being hit by light flak, to whom Leigh-Mallory offered a lift back to RAF Thorney Island in his personal Douglas DC-3 Dakota.
(IWM CL129)

Had it not been for the comprehensive interdiction programme by the RAF and USAAF before D-Day and during the beachhead battles, and the air bombardment delivered at the right place and the right time, there was no way the Allied land forces could have gained a toe-hold on the beaches, held their bridgeheads against fierce enemy counter-attacks and then made their breakout towards the Seine and Paris.

The two Tactical Air Forces – the 2nd (British, 2nd TAF) and the 9th (US, 9th AF) – were formed in 1943 to give direct air support to the British and American land forces during and after the D-Day landings. The formation of 2nd TAF was informed by the RAF's experience in North Africa with the 8th Army and the Desert Air Force, where fighters and bombers operated in close support of the ground forces, all working together in a joint plan. The US 9th AF came to Europe straight from the USA in 1943 and was expected to work up to combat readiness for the invasion.

The RAF began practical preparations in the summer of 1943 for its role in the forthcoming invasion, when ground staff and the base organisation of a number of tactical units were drastically reorganised. A series of numbered 'airfields' was established, with pairs of airfields

forming sectors. These airfields were fully mobile and were able to operate efficiently for indefinite periods of time from forward airstrips (or landing grounds). Ground crews were detached from their squadrons and formed into independent servicing echelons, which were then attached to a particular airfield to attend to the needs of any squadrons and their aircraft that landed there. Certain squadrons soon became identified with particular airfields, wings and ground crew servicing echelons, and they moved onwards to different airfields together. This practice was honed in England during 1943 and into 1944, so that when the time came for the airfields and servicing echelons to move to their new operating bases at the advanced landing grounds in Normandy they had become fully proficient in their roles.

Until aircraft landing grounds could be established in France, the fighter-bomber squadrons of the 2nd TAF and 9th AF flew tactical air-support missions across the Channel from temporary Advanced Landing Grounds (ALG) in the southern counties of Kent, Sussex and Hampshire; and in France and further afield from 6 June 1944 up until the end of the war in Europe on 8 May 1945.

Air Chief Marshal Sir Trafford Leigh-Mallory, Commander-in-Chief of the Allied Expeditionary Air Forces (AEAF) was the first to propose the building of these temporary airfields and it was one of the most important tasks given to the RAF Airfield Construction Service (RAF ACS) prior to D-Day. These ALGs differed

from the dozens of permanent airfields built across central and eastern England since 1939 to sustain the strategic air offensive against Germany by the RAF and USAAF. They were tactical combat airfields; temporary affairs to be used by the tactical air forces in support of the advancing ground forces on the battlefield across the Channel.

Sited on requisitioned farmland, each ALG had two metal track runways that measured 1,600yd by 50yd, made with Sommerfeld Track, Square Mesh Track (SMT) or Pierced Steel Plank (PSP), with 2½ miles of perimeter track and an additional MT track running parallel with the perimeter. Accommodation was sparse and rudimentary, with existing farm buildings often put to use. Up to four blister hangars were also provided.

About 12 ALGs became active in southern England during the summer and autumn of 1943, but the squadrons and their aircraft moved to permanent bases over the wet winter months. When spring 1944 arrived and the ground became dry enough, the squadrons returned to the ALGs, from where they resumed operations against tactical targets in northern France and Belgium.

ABOVE **Two Typhoon Mk IBs of 174 Squadron raise the dust as they take off from B5/Le Fresne-Camilly on 13 July 1944.** *(IWM CL450)*

Advanced Landing Grounds in France

Fighter-bombers based at the ALGs in southern England could spend only a short time over the Normandy battlefront owing to the distance they had to fly across the Channel, and the need for them to return again to base to refuel and rearm. It was estimated that 100 miles from base was the effective limit of their operational use.

However, the destruction of enemy airfields and their facilities in northern France by 2nd TAF and 9th AF had been necessary to ensure Allied air supremacy over the invasion area, but it came at a price. Usable runways and maintenance facilities in Normandy were denied to the Allied air forces, which meant they would need to build their own airfields from scratch on mainland Europe. So the air forces deemed it a priority that temporary landing grounds should be constructed in Normandy (especially on the

ABOVE **Maintaining aircraft on a USAAF ALG.** *(USNA)*

Caen–Falaise plain, which air force planners considered particularly suitable) as soon as possible after the beachhead landings.

They also recognised that once the Allied front line had been pushed forward and beyond the effective range of the fighter-bombers, the wings, squadrons and servicing echelons would need to move to newly built landing grounds that were closer to the action, leaving the ones in the rear for supply and medical evacuation purposes (when the time came, some ALGs were simply abandoned).

These landing grounds were divided into four types, for which the construction programme priorities were established as follows:

Priority I: Emergency Landing Strips (ELS)
Priority II: Refuelling and Rearming Strips
              (RRS)
Priority III: Advanced Landing Grounds (ALG)
              (that would later become airfields)
Priority IV: Airfields – with the same facilities
              as an ALG, although with improved
              dispersal facilities.

Emergency Landing Strips (ELS) – A
minimum length of 1,800ft, built on flat ground that had been roughly graded. As the name implies, these strips were not suitable for routine operations, but to enable pilots in difficulty to make an emergency landing.

Refuelling and Rearming Strips (RRS)
– A minimum length of 3,600ft with a level compacted surface suitable for take-offs and landings, plus two marshalling areas (each 50yd by 100yd at both ends of the runway) to allow fast turnaround of aircraft, and adequate tracking to enable operations to continue under all normal summer and autumn weather conditions.

Advanced Landing Grounds (ALG) – A
minimum length of 3,600ft for fighters, and 5,000ft for fighter-bombers, with the same facilities as a RRS and dispersals for 54 aircraft of both types, and the capability to be used to full capacity.

Airfields – With the same facilities as an ALG, although with improved dispersal facilities.

The sites
Potential sites for landing strips had been selected by Allied planning staff in advance of the invasion, based on expert knowledge of the geology of Normandy. The plateau of Calvados around Bayeux and Caen, where the RAF ALGs were to be sited, possessed the requisite qualities in terms of the substrata and surface deposits that provided excellent drainage in wet weather; a firm and even surface in dry weather that was not prone to cracking; and large open

BELOW Work continues on building this landing strip as a P-47 Thunderbolt takes off after refuelling and rearming. *(USNA)*

fields of arable soil of consistent quality, without hedges or ditches.

For the American airfields that were to be sited mainly in the Cotentin peninsula, the ground was more challenging for the IX Engineer Command construction battalions. Much of it was clay soil, which meant it would require a greater proportion of surface tracking.

Initial plans called for the following minimum programme for airfield construction in Normandy:

- 3 ELSs (2 American, 1 British) by the end of D-Day.
- 4 RRSs (2 American, 1 British) by evening on D+3 and not later than D+4.
- 10 ALGs (5 American, 5 British) by D+8 (included the four RRSs).
- 18 Airfields (8 American, 10 British) by D+14.
- 27 Airfields (12 American, 15 British) by D+24.
- 43 Airfields (18 American, 25 British) by D+40.
- 93 Airfields (48 American, 45 British) by D+90.

Such was the importance attached to the rapid construction of forward airfields in Normandy that advance parties from the British and US Airfield Construction Groups were landed on D-Day, with the main bodies of the units, their equipment and plant following over the next few days. Once the armies were ashore, airfield construction took precedence over all other construction activities until they were built.

Building the airfields

To establish a firm runway surface that was able to sustain constant use by heavily loaded fighter-bomber aircraft, the topsoil was removed by scraper and the subsoil was compacted at optimum moisture content using sheep's foot and wobble wheel rollers. As predicted, the loessic soils beneath the British airfields drained rapidly in wet weather. However, the combination of fine weather with the agitation of fine-grain loessic soils produced dense clouds of dust during the airfields' early operational use. Some airfields were sprayed with water to dampen down the dust in an attempt to

ABOVE Australian pilots of 453 Squadron help to flatten the airstrip at B11/ Longues-sur-Mer. Flg Off D. Osborne and Plt Off A. Rice man a Jeep, while Plt Off J. Scott steadies the two 500lb bombs being used to add weight to a locally acquired agricultural roller. In the background, Spitfires of 602 Squadron depart on an operation, watched by a runway controller working from a converted sentry box. (IWM CL509)

minimise the hazard to visibility and reduce engine wear.

Where possible, the airfields were constructed with minimum disturbance to any grass surface (and accepting a considerable roughness in dispersal areas), but the only real solution to the dust problem was to operate from surfaced runways.

The RAF's first ELS (B1/Asnelles) was completed and ready for use on 7 June (D+1). Five more airfields were finished by 17 June (D+11), and by the end of the month the RAF

BELOW Typhoon Mk IB, MN529 'BR-N', of 184 Squadron RAF raises clouds of dust as it takes off from B2/ Bazenville on a sortie armed with rocket projectiles, 14 June 1944. (IWM CL147)

ABOVE Aerial view of A1/St-Pierre-du-Mont constructed by the US IX Engineer Command's 834th Engineer Aviation Battalion (EAB). *(USNA)*

were making use of 10 operational airfields in Normandy.

Within 24 hours of D-Day the USAAF's IX Engineer Command had set up two airfields – ELS1/Poupeville and A21/Saint-Laurent-sur-Mer. Its men worked in difficult conditions, often having to set their tools aside to deal with sniper fire.

Facilities on ALGs were fairly basic, with personnel usually sleeping in tents and flying control operating from another tent. Aircraft were often serviced in the open, but when work needed to be carried out under cover, prefabricated steel and canvas Butler hangars came in useful.

Manpower

The work in Normandy was undertaken jointly between the RAF and USAAF – one RAF Airfield Construction Wing (RAF ACW) and five Royal Engineers Airfield Construction Groups, which had trained in earnest for airfield building from the summer of 1942; and 17 Engineer Aviation Battalions (EABs) of the US 9th Air Force's IX Engineer Command, which was formed to build and maintain airfields in England and on the European continent for the US 9th Air Force in support of the Allied forces.

RAF ACW and Royal Engineers

An RAF Airfield Construction Wing was made up of four squadrons, each of which comprised six Field Flights, a Plant Flight and an MT Flight. A Royal Engineers group was made up of a Headquarters Company and two Road Construction Companies, with each of the latter having a Pioneer Company attached and under operational command. The plant they used included crawler tractors, dump trucks, motor graders, rollers, scrapers and transporters.

Five RAF Airfield Construction Wings subsequently landed on the Continent: 5357 Wing's squadrons worked with the front-line forces, building and repairing airfields for the fighter and fighter-bomber squadrons as they advanced; the other four wings undertook wider-ranging programmes of essential construction and repair work in support of both Army and RAF operations.

RIGHT A USAAF P-38 Lightning lands at Emergency Landing Strip 1 (ELS-1) raising a dust cloud in its wake. In the right foreground is a Piper L4 Grasshopper, which was used for a variety of roles in Normandy, including artillery spotting and medical evacuation. *(USNA)*

LEFT **RAF ACS and Royal Engineers lay Square Mesh Track (SMT) at B3/Saint-Croix-sur-Mer.** *(USNA)*

BELOW Road building in Normandy was also undertaken by British airfield construction teams. *(USNA)*

ABOVE An RAF grader and its driver. *(USNA)*

US IX Engineer Command

The basic work unit of the IX Engineer Command was the Engineer Aviation Battalion, which was assigned works projects that were dealt with as its commander saw fit – the end result was all his superiors were interested in. A battalion could be assigned entirely to construction work, or could be detailed for maintenance work, or both. A total of 16,000 officers and men made up the 17 EABs. Because their officers were specialist engineers and the men were experienced from time spent building airfields in the Pacific theatre, they needed little further training to equip them for the task that lay ahead in Normandy.

RIGHT SMT is laid on a partly completed ALG. *(USNA)*

Here is an outline of the plan devised by the IX Engineer Command to streamline the construction of its ALGs in Normandy:

Personnel and equipment in a battalion were divided into echelons, A, B, C and D. The rationale behind this was to devise a priority of movement that would bring to the airfield construction site during each phase of operations the manpower and equipment of most use to the operations at the particular time.

The first echelon to reach the planned airfield construction site was the reconnaissance party, which would get there as quickly as the prevailing tactical situation permitted. Here they would look over the location and decide upon the best runway site and dispersal areas, examine the area for enemy mines, and carry out whatever preliminary layout could be done in the time available.

Next to arrive was the advance party led by the battalion commander. It would make contact with the reconnaissance party, and the battalion commander would make his decision as to whether the site was suitable or not, and whether the proposed layout of the

reconnaissance party should be followed, or whether it should be changed. If he decided to construct on the site, preliminary construction could get under way; if he decided that the site was unsuitable, the reconnaissance party would set out for an alternative site where the process would be repeated.

The advance party comprised 50 men, some of which were to assist in the further layout work, and others to protect the working party. For equipment, the echelon had two D-7 tractors, two motorised graders and a truck, so initial earth moving and clearing could begin.

The next echelon to arrive was the support echelon, which would add to the clearing and earth moving with two D-7 tractors with blades and carryall scrapers, one D-4 tractor, and two petrol saws, plus eight trucks with which to start transporting construction materials to the site. This echelon would add 100 men to the team, for whatever use.

By the time the main body arrived with 541 men, the project would be sufficiently 'opened up' to use the rest of the earth moving and clearing equipment, and perhaps the track laying could begin.

The main body brought with it all the remaining equipment considered necessary to sustain rapid construction and finish the job, including four more D-7 tractors with blades and carryall scrapers, two D-4 tractors, four motorised graders, 28 2½-ton trucks, two air compressors, a truck-mounted crane, a tractor mounted crane, a trailer-mounted lubricator, four 16-ton trailers, various rollers, four petrol saws, water supply equipment and 13 administrative vehicles.

Runway construction and surfacing materials

Facilities on a landing ground were to be provided in the following order:

- Runway (tracked, if necessary).
- Taxi-tracks and hardstands or dispersal areas.
- Development and operation of water points.
- Improvement and/or construction of access roads.
- Protection of Flying Control installation.
- Access roads to storage areas.
- Erection of petrol storage areas.
- Minimum rehabilitation of damaged buildings in the vicinity for Headquarters and weather-sensitive supplies.

Instead of using rough, unimproved dirt landing strips, engineers used a range of specially developed surfacing materials – Sommerfeld Track, Square Mesh Track, Pierced Steel Plank and Prefabricated Hessian Surfacing – to strengthen the soil and support the weight of an aircraft, and as a measure of insurance against the inevitable wet weather. Initially, the airfields were single-runway landing strips, laid down east–west (09/27) unless local conditions dictated a different runway orientation.

Sommerfeld Track

ALGs laid in the UK prior to D-Day were surfaced with Sommerfeld Track, a lightweight flexible wire mesh type of prefabricated airfield surface named after the expatriate German-Jewish engineer, Kurt Joachim Sommerfeld, who fled to England in 1933 to escape persecution by the Nazis. First used by the British in 1941 as road and runway surfacing, Sommerfeld Track was flexible in a longitudinal direction but stiff or rigid laterally. It consisted of a fabricated surface comprising a wire netting stiffened laterally by steel rods threaded through at 9in intervals. This gave it a load-carrying capability while staying flexible enough to be rolled up. Nicknamed 'tin lino', Sommerfeld Track came in rolls measuring 10ft 8in wide by 75ft 6in long, which could be joined at the edges by threading a flat steel bar through loops in the ends of the rods.

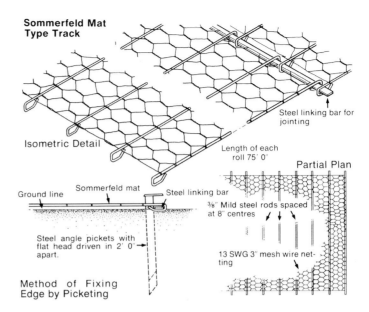

Sommerfeld Mat Type Track

Isometric Detail

Ground line   Sommerfeld mat   Steel linking bar

Steel angle pickets with flat head driven in 2' 0" apart.

Method of Fixing Edge by Picketing

Steel linking bar for jointing

Length of each roll 75' 0"

Partial Plan

³⁄₈" Mild steel rods spaced at 8" centres

13 SWG 3" mesh wire netting

Square Mesh Track (SMT)

The surfacing material selected for the building of ALGs during the first weeks after the Normandy invasion was known as Square Mesh Track (SMT). SMT was developed by the British Reinforced Concrete Engineering Co. in 1943 and made from heavy wire joined in 3in squares. It was chosen over other surfacing materials because it was very lightweight, which allowed sufficient quantities to be transported across the English Channel on the hard-worked landing craft. SMT was an easily managed material, and a landing ground for fighters could be laid like a carpet in about one week.

ABOVE Sommerfeld Track. *(Author's collection)*

LEFT An engineer makes a repair to the SMT surfacing. *(USNA)*

1   The airfield at A9/Le Molay, situated about 10 miles south-west of Bayeux, under construction by men of the 834th EAB. *(USNA)*

2   Irrigation ditches that run the width of the runway have already been partially filled by bulldozers. The road scraper, towed by a heavy tractor, carts earth from a nearby location and lays it evenly along the ditch, completing the fill. The road scraper, also commonly known as the carryall or 'pan', can carry 8cu yd of earth and perform the dual job of collecting excess soil, or depositing it where needed. *(USNA)*

3   Smooth-wheel rollers run the length of the runway again and again, gradually packing the surface until it is smooth and hard. These rollers weigh 10 tons each and are responsible for the finished surface that aircraft need for a good landing. *(USNA)*

4   A dump truck empties earth into an uneven place on the runway at A1/Saint-Pierre-du-Mont a few days after D-Day. *(USNA)*

5 After the scraping and scarifying processes by the graders, the water tankers wet down the broken-up sod prior to packing it down. These tankers will continue to operate throughout the entire construction of the runway, wetting it down in advance of each packing-down process. They will be the last to leave the completed airfield, after a final wetting down to settle the dust. *(USNA)*

6 A crane is used for loading Prefabricated Bituminised Surfacing (or hessian mat) on to a truck prior to its laying. *(USNA)*

7–10 After the selected site for the airstrip has been graded and levelled, hessian mat (which has a tar-paper-like composition) is laid by lorries loaded with 200ft rolls. A 50% overlap is maintained with adjacent matting, which is then cemented together with a solution of diesel oil mixed with petrol. The mat is rolled with a wobble-wheel roller and then sealed with a jet of liquid cement from a dispenser on another lorry. *(USNA)*

RIGHT Laying Prefabricated Bituminised Surfacing (PBS). (USNA)

Prefabricated Bituminised Surfacing (PBS) or Prefabricated Hessian Surfacing (PHS)

After the initial batch of Normandy airfields was completed using SMT, the construction engineers switched almost exclusively to another surfacing material known as Prefabricated Bituminised (burlap) Surfacing (or PBS). Developed by Canadian Army Engineers from British origins as an aid to rapid surfacing, PBS was light and easily transportable, with good all-weather qualities. It could be laid quickly and did not create the dust problems encountered with SMT-surfaced landing grounds. Trials of PBS held in the summer of 1943 were a success, and led to large quantities of the surfacing material being manufactured. By the war's end some 20 million square yards had been produced in the UK alone.

Simply described as a hessian or jute impregnated cloth, it was delivered in 300ft-long rolls, 36in or 43in wide, and laid in overlapping layers sealed together at the edges to produce a dust-free fair-weather surface.

PBS was a material that needed careful handling and required regular maintenance in use. It was quickly learned that harsh braking action by aircraft caused PHS to tear, but this was remedied by the practice of laying SMT over the top of PHS surfacing in vulnerable locations. But the main attraction of PBS was the speed with which it could be laid. A team of 150 men could surface a 3,600ft by 120ft runway in about 14 hours.

PBS was also a cost-efficient alternative to PSP – an average ALG required some 800 tons of PBS against 4,800 tons of PSP for an equivalent landing site. There was also huge saving in shipping space in the LSTs that carried the cargo across the Channel.

Pierced Steel Plank (PSP)

To provide an all-year-round airfield for the US 9th Air Force's medium and light bombers, a third type of surfacing material known as Pierced Steel Plank (PSP, or Marsden Matting) was introduced to Normandy in July 1944. It consisted of 10ft-long by 15in-wide perforated steel planks joined together and laid perpendicular to the line of flight. Used widely in other theatres of operations, PSP was an ideal surfacing for all airfields on the Continent, but its limited availability and greater weight made this impractical. Moreover, because of supply problems, the construction of even one PSP-surfaced fighter-bomber airfield could take a month or longer, while similar PBS and SMT fields could be constructed in two weeks and one week, respectively.

Sod and earth

In addition, sod and earth runways were built for ELSs and RRSs.

BELOW Men of the 833rd EAB lay PSP during the construction of an airfield. (USNA)

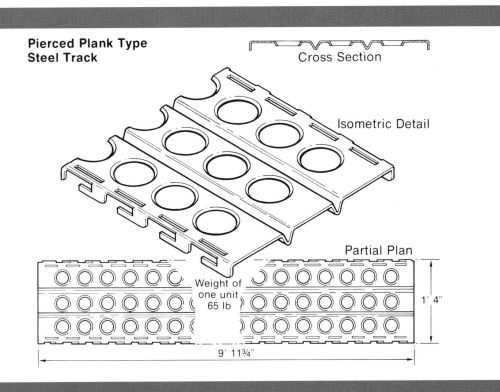

Pierced Plank Type Steel Track

Cross Section

Isometric Detail

Partial Plan

Weight of one unit 65 lb

1' 4"

9' 11¾"

American ALGs used from 6 June to 24 July 1944

Code no/name	Initial operational date
ELS1/Poupeville	7 Jun
A1/St-Pierre-du-Mont	13 Jun
A2/Criqueville	19 Jun
A3/Cardonville	14 Jun
A4/Deux-Jumeaux	30 Jun
A5/Chippelle	5 Jul
A6/Beuzeville	15 Jun
A9/Le Molay	30 Jun
A11/St Lambert	5 Aug
A12/Lignerolles	18 Jul
A13/Tour-en-Bessin	28 Jul
A21/St-Laurent-sur-Mer	7 Jun
A22/Colleville-sur-Mer	13 Jul

Construction materials received by IX Engineer Command (in long tons) 6 June to 9 August 1944

PSP	13,115
PHS	5,264
SMT	12,242

British ALGs used from 6 June to 8 August 1944

Code no/name	Initial operational date
B1/Asnelles ELS	7 Jun
B2/Bazenville	11 Jun
B3/St-Croix-sur-Mer	10 Jun
B4/Beny-sur-Mer	15 Jun
B5/Le Fresne-Camilly	15 Jun
B6/Coulombs	15 Jun
B7/Rucqueville	Unknown
B8/Sommervieu	22 Jun
B9/Lantheuil	22 Jun
B10/Plumetot	10 Jun
B11/Longues-sur-Mer	21 Jun
B12/Ellon	18 Jul
B14/Amblie	7 Jul
B15/Ryes	5 Jul
B16/Villons-les-Buissons	7 Aug
B17/Carpiquet	8 Aug
B18/Cristot	25 Jul
B19/Lingevres	6 Aug
B21/Ste-Honorine-de-Ducy	8 Aug

Airfield construction American-style

The US 843rd Engineer Aviation Battalion arrived in Normandy on 2 July. Their first assigned mission was to construct an Advanced Landing Ground (A16) at Brucheville (Le Grosellier) near Carentan. The battalion's diarist wrote about their experiences over one month of battling the weather to mark out and clear the ground before building an airfield on Norman soil:

We pulled out of Saint-Lambert at 16:00hrs on 5 July, going by motor convoy along the main trunk road which ran south from Cherbourg, a road jammed with trucks hauling men and supplies, through Carentan which at that time was under sporadic enemy artillery fire, past infantry and artillery positions dug in along an airstrip occupied by our fighter planes, on to Le Grosellier (Brucheville), where we were to build our first complete ALG in France.

ABOVE Graders with their blades lowered make the first cut. Following the lines of stakes set by the surveyors, they start levelling the runway area. Later they raise their blades and lower a set of large teeth located in the front of the machines, and begin to scarify, or break up, the ground. *(USNA)*

ABOVE SMT is laid by men of the 833rd EAB. *(USNA)*

At this time the Carentan sector of the Cherbourg peninsula was a hotly contested area, heavy rains in an already marshy lowland country having made the going even more difficult. In the marshland corridors through which attack was possible, enemy infantry had dug positions in hedgerows and was supported by heavy mortar fire, machine guns, and an increasing amount of artillery. Planes as well as heavy artillery were needed to blast the enemy out of these strong positions. It was our immediate task to construct landing strips to increase the effectiveness of air support for our ground forces fighting now across a 10-mile front in this southwest sector of the Cherbourg peninsula.

The Brucheville mission had been assigned the 843rd on 4 July 1944. The specifications called for one runway, 5,000ft x 120ft; 3,600ft to be surfaced with PHS and the remainder graded and compacted earth; one taxi track surfaced with PHS and one with Square Mesh Track; 75 hardstands, 42ft x 72ft, 50 to be surfaced with mesh track and 25 to be graded and compacted only. The completion date was set for 16 July 1944.

An advance party of the Battalion S-3 section had moved to the new location on the morning of the 5th and staked the centre-line of the runway. On 6 July, clearing of hedgerows from the runway site was started with equipment working three shifts of six hours each. All trees and brush were removed and the soil was stripped to a depth sufficient to reach a satisfactory base, backfilled and compacted with sheep's foot rollers. Clearing of hedgerows was completed on 9 July. Grading of the runway started on 8 July and progressed rapidly until on 13 July a large soft spot was encountered on the heavy cut at a point along the middle of the runway. It was necessary to excavate all material from this spot to an average depth of 4ft, install drain pipe to remove the subsurface water and backfill with rock from a pit opened near the field. Clearing of taxiways was started on 12 July to utilise equipment not needed on runway grading. Clearing of approach funnels of the runway was completed the following day, and on this date laying of hessian matting on the runway was started with two companies working two eight-hour shifts.

The PHS, or Prefabricated Hessian Surfacing, was a hessian cloth heavily coated on both sides with bitumen. After the ground was compacted at an optimum moisture content to form a hard, dense surface, this

ABOVE Carpenters go to work with pneumatic tools to construct the control tower for an airfield. The compressor to the left provides the power for the pneumatic saws, which cut the timber quickly and cleanly. Six carpenters were able to finish this tower within two hours of the time this picture was taken. (USNA)

ABOVE Aircraft on this landing strip (either A9/Le Molay or A7/Azeville) are directed for take-off and landing by two American soldiers from this checkerboard-covered cart: Sgt Charles L. Kautz (left) and Sgt Francis B. Boone. In the background can be seen an F-6A (the reconnaissance version of the P-51A Mustang) of the USAAF's 107th Tactical Reconnaissance Squadron, the first recce unit to operate from French soil after D-Day. (USNA)

roofing material was laid, each sheet with a 50% lap to give a two-ply covering. The bitumen on one side of the material was dipped into a mixture of gasoline and diesel oil just before laying; this softened the bitumen coating until it became very sticky, and thus provided its own adhesive. Canadian Army Engineers are credited with developing the idea, but the material that we used had been manufactured by American roofing material manufacturers. The cloth was 43in wide and each roll was 300ft in length. Usually the area at the ends of the runway subjected to turning action of the planes had a layer of wire mesh laid on top of the hessian to prevent tearing. The hessian material went down rapidly and an entire runway could be laid in three days, but the base had to be perfect – and it could not be laid effectively in the rain. This covering was very successful – no dust, no mud, and the pilots liked it.

On 14 July, a second soft spot was encountered on the runway, causing more delay in the grading operations, and as a result changed the completion date to 21 July. On 15 July grading and installation of taxi-track surfacing was started. Heavy rains delayed laying of hessian matting for two days and also made it necessary to remove and re-lay some of the mat which was not sealed at the edges when the rain started. Heavy rains, commencing on the afternoon of 20 July, again stopped all operations except drainage for three days. This delay moved the completion date back to 1 August. On 24 July, materials for two 250-barrel

bolted steel tanks was received and erection started. Fair weather allowed good progress for three days, but then rain again slowed grading operations and delayed mat-laying. The field was made operational with the runway, one taxiway and 38 hardstands complete on 1 August. On 2 August the first planes flew in from England and landed on the field. Work continued on one taxiway, access roads and hardstands, and the field was completed on 7 August.

ABOVE This steel 10,000gal petrol storage tank could be dismantled and reassembled at its next airfield by a team of six men in about three days. (USNA)

Airfield construction British-style

The RAF's 5023 Airfield Construction Squadron landed in Normandy at Le Hamel at 3:00pm on 6 July and by the evening of the 8th, complete with personnel, plant, MT and equipment, they were ready for work. The squadron history describes what happened next:

For the first few days our main work consisted of maintaining the airstrips, construction of roads, perimeter tracks, making cannon butts, oiling the strips to keep down the dust, and general minor works. Our plant operators had their first experience of working under shellfire whilst assisting the Royal Engineers to build a new road at Tilly. All the squadron had its first experience of digging in so as to be safe from air attacks, which happened almost nightly. Soon we began to talk about 'strips', not airfields, and we referred to them by letters and numbers, B2, B3, B4, B7, B8, B9 – these strips made famous by the flying aces flying from them, soon took on the appearance of 'pukka airfields' – we worked on them all from dawn to dusk – no rest – no complaints

ABOVE Plant operators of 5023 Airfield Construction Squadron, RAF. *(Author's collection)*

BELOW RAF flying control personnel at work at B2/Bazenville, five miles north-east of Bayeux, on 15 June. *(IWM CL162)*

– this is what we had been trained for and we were fit, happy and proud to be a part of the Second Front.

On 31 July we moved three flights to a site near Tilly-sur-Seulles, to be followed by the move of the remainder of the squadron on 2 August. Here we commenced work with our sister squadron, 5022, to construct an advanced landing ground. Never has an Airfield Construction Wing or Group accomplished so much in such a short time. When we arrived, the corn was growing to a height of 3–4ft, the trees were in full foliage and looking very permanent. In five days a startling change took place. A Square Mesh Track runway 1,300yd long x 40yd wide with SMT taxi track 40ft wide surrounding it, and a 50ft-wide MT road circling the airfield and serving five dispersals were all completed – B19, a new airfield, was ready for operational flying. As we rested on 8 August to shave for the first time for days, we realised that we had surprised even ourselves.

LEFT Airmen of 419 Repair and Salvage Unit, aided by an AEC mobile crane, remove damaged Typhoon Mk IB, MN413 'I8-T', of 440 Squadron RCAF from the PSP landing strip following a wheels-up landing at B9/Lantheuil (eight miles east of Bayeux) on 1 August. *(IWM CL652)*

RIGHT Spitfire Mk IXs of 421 Squadron RCAF prepare to taxi out from their dispersals at B2/Bazenville for a dusk patrol on 13 August. *(IWM CL782)*

Epilogue
The butcher's bill

When reading about the science and machines that helped make D-Day possible it is easy to lose sight of the fact that the invasion was a human endeavour. Large numbers of real people were involved; real blood was shed and real lives were lost.

With the precision that went into planning Neptune and Overlord it may come as a surprise to learn that there is no 'official' casualty figure for D-Day. In the fog of war, accurate record keeping was difficult. For example, troops who were listed as missing may actually have landed in the wrong sector and rejoined their parent unit only later.

The Allied casualty figures for D-Day, 6 June, have generally been estimated at around 12,000 to 14,000, including 2,500 dead. This figure can be broken down by nationality as follows: (approximately) 2,700 British, 946 Canadian, and 6,603 American. Recent research by the US National D-Day Memorial Foundation has reached a more accurate – and much higher – figure for Allied personnel killed on D-Day, verifying 2,499 American D-Day fatalities and 1,915 from the other Allied nations, giving a total of 4,414 dead (far higher than the hitherto accepted figure of 2,500 dead). Further

ABOVE The Normandy American Cemetery and Memorial at Saint-Laurent-sur-Mer is situated on a cliff overlooking Omaha Beach. It contains the graves of 9,387 US servicemen and women, many of whom died on Omaha. *(Shutterstock. com/Jakez)*

research could mean that these numbers will increase.

Casualties on the British beaches numbered approximately 1,000 on Gold Beach and the same again

on Sword Beach. The remainder of the British losses were suffered among the airborne troops: some 600 were killed or wounded, and 600 more were missing; 100 glider pilots also became casualties. The losses of the 3rd Canadian Division at Juno Beach have been given as 340 killed, 574 wounded and 47 taken prisoner of war (not including the British engineer and commando units that landed on Juno).

US casualties were as follows: 1,465 dead, 3,184 wounded, 1,928 missing and 26 captured. Of the total US figure, the US airborne troops suffered 2,499 casualties, of which 238 were killed. By comparison, the casualties at Utah Beach were light: 197, including 60 missing. However, at Omaha Beach the US 1st and 29th Divisions together suffered at least 2,000 casualties.

The precise number of German casualties on D-Day is not known, but it is likely to have been somewhere between 4,000 and 9,000 men.

More than 425,000 Allied and German troops were killed, wounded or were listed as missing during the Battle of Normandy (6 June until 31 August). This figure includes over 209,000 Allied casualties, of which nearly 37,000 dead were from ground forces and 16,714

ABOVE La Cambe German Cemetery near Bayeux contains in excess of 21,000 German war dead, including many from the Normandy campaign. *(US Army Photo/Alfredo Barraza Jr)*

deaths from the Allied air forces. Among the Allied casualties, 83,045 were from 21st Army Group (British, Canadian and Polish ground forces) and 125,847 from US ground forces.

The exact figure of German losses in the Battle of Normandy can only be guessed at. Some 200,000 German troops were killed or wounded, while the Allies took 200,000 prisoners of war (which are not included in the 425,000 total, above). During the heavy fighting around the Falaise Pocket in August the Germans lost about 90,000 men, including those taken prisoner.

In Normandy today there are 27 war cemeteries that hold the remains of over 110,000 dead from both sides: 77,866 German, 9,386 American, 17,769 British, 5,002 Canadian and 650 Poles.

However, combatants were not the only casualties in the battle. Some 14,000 French civilians were killed between 6 June and 21 August in the three *départements* of Lower Normandy where the fighting took place, mainly due to Allied bombing, and at least 250,000 more (some estimates reckon as many as 400,000) fled their homes in town and country to escape the fighting.

LEFT Beny-sur-Mer Canadian War Cemetery contains 2,048 Second World War burials, most of them men of the 3rd Canadian Division who died on D-Day and in the early days of the advance towards Caen. *(Burtonpe/Wikimedia Commons)*

Sources

PRIMARY

The National Archives (UK)

AIR 27/717 – Operations Record Book, 88 Squadron RAF, Jan 1944–Apr 1945

AIR 27/802–803 – Operations Record Book, 101 Squadron RAF, Jan–Dec 1944

AIR 27/966–968 – Operations Record Book, 140 Squadron RAF, May 1941–Nov 1945

AIR 27/1004 – Operations Record Book, 149 Squadron RAF, Jan 1944–May 1945

AIR 41/24 – RAF Narrative: The Liberation of North West Europe, Vol. III: The Landings in Normandy

DEFE 2/416–417 – Report by Naval Commander, Force 'G', Operation Neptune, 6 June 1944

WO 291/1331 – No 2 Operational Research Section, 21st Army Group

SECONDARY

Administrative History of US Naval Forces in Europe, 1940–1946, Vol 5. *The Invasion of Normandy: Operation Neptune* (Commander US Naval Forces in Europe, 1946, Washington DC)

Anderson, Richard C. Jr., *Cracking Hitler's Atlantic Wall: The 1st Assault Brigade Royal Engineers on D-Day* (USA, Stackpole Books, 2010)

Anon, *An Account of The Operations of Second Army in Europe 1944–1945*, Vol I (Second Army Headquarters, repr. Military Library Research Service, 2005)

Anon, Battle Summary No. 39: *Operation Neptune: Landings in Normandy, June 1944* (London, HMSO, 1994)

Anon, *Omaha Beachhead (6 June – 13 June 1944),* 'American Forces in Action' Series (US War Department, Historical Division, 1945)

Anon, *United States Forces, European Theater: Armored Special Equipment* (Office of the Chief of Military History, General Reference Branch, US Army, Washington DC, 1953)

Bailey, Roderick, *Forgotten Voices of D-Day* (London, Ebury, 2009)

Balkoski, Joseph, *Omaha Beach: D-Day, June 6, 1944* (USA, Stackpole Books, 2004)

Balkoski, Joseph, *Beyond the Beachhead: the 29th Infantry Division in Normandy* (USA, Stackpole Books, 2005)

Barbier, Mary Kathryn, *D-Day Deception: Operation Fortitude and the Normandy Invasion* (USA, Stackpole Books, 2009)

Brown, D.K., *The Design and Construction of British Warships 1939–1945, The Official Record: Landing Craft and Auxiliary Vessels* (London, Conway Maritime Press, 1996)

Clough, Brigadier A.B., *Maps and Survey* (London, HMSO, 1952)

Fletcher, David, *Swimming Shermans: Sherman DD amphibious tank of World War II* (Oxford, Osprey, 2006)

Fowle, Barry W. (Ed), *Builders and Fighters: US Army Engineers in World War II* (USA, Office of History, US Army Corps of Engineers, Fort Belvoir, Virginia, 1992)

Friedman, Norman, *US Amphibious Ships and Craft: An Illustrated Design History* (Annapolis, USA, Naval Institute Press, 2002)

Gawne, Jonathan, *Spearheading D-Day: American Special Units of the Normandy Invasion* (Paris, Histoire et Collections, 2001)

Gordon, Alan, 'Mapping and charting for the greatest collaborative project ever' (*The American Surveyor,* 2005)

Jarman, W.D., *Those Wallowing Beauties: The Story of Landing Barges in World War II* (Lewes, The Book Guild, 1997)

Keller, Maj-Gen R.F.L., 'The Technique of the Assault: the Canadian Army on D-Day: After-action reports by commanders' (*Canadian Military History*, Vol 14, No 3, Summer 2005)

Lavery, Brian, *Assault Landing Craft: Design, Construction & Operations* (Barnsley, Seaforth Publishing, 2009)

Lewis, Adrian R., *Omaha Beach: A Flawed Victory* (USA, University of North Carolina Press, 2001)

Monckton, Sir Walter, Report: 'The part played in "Overlord" by the synthetic harbours' (Chiefs of Staff Committee, 18 January 1946)

Pakenham-Walsh, Maj-General R.P., *Military Engineering (Field)*, (The Second World War 1939–1945, Army), (London, The War Office, 1951)

Pemberton, Brigadier A.L., *The Development of Artillery Tactics and Equipment* (The Second World War 1939–1945, Army), (London, The War Office, 1950)

Prados, Edward F. (Ed), *Neptunus Rex: Naval Stories of the Normandy Invasion, June 6, 1944* (USA, Presidio, 1998)

Price, Alfred, 'Spoof operations' (*Aeroplane* magazine, June 2004)

Rose, Edward P.F., and Pareyn, Claud, 'British applications of military geology for Operation Overlord and the battle on Normandy, France, 1944' (Geological Society of America, *Reviews in Engineering Geology*, Vol XIII, 1998)

Roskill, Captain S.W., *The War at Sea 1939–1945, Vol III: The Offensive, Part II, 1st June 1944–14th August 1945* (London, HMSO, 1961)

Sailplane and Glider magazine, 'Tank carrying glider', Vol 13, No 2, March 1945

Shaw, Frank and Joan (compilers), *We Remember D-Day* (Hinckley, 1994)

Skill in the Surf: A landing boat manual (US Navy, February 1945)

Smith, David J., *Britain's Military Airfields 1939–45* (Wellingborough, PSL, 1989)

The Civil Engineer in War: a symposium of papers on wartime engineering problems, Vol 2: 'Docks and Harbours' (The Institution of Civil Engineers, 1948, London)

Trew, Simon (Ed), *Battle Zone Normandy*, various volumes (Stroud, Sutton Publishing, 2004)

Trew, Simon, *D-Day and the Battle of Normandy: a photographic history* (Sparkford, Haynes, 2012)

US Air Force Doctrine Document 2-5.1. 'Electronic Warfare' (USAF, November 2002)

Wood, Alan, *History of the World's Glider Forces* (Wellingborough, PSL, 1990)

Zaloga, Steven J., *Rangers Lead the Way: Pointe-du-Hoc, D-Day 1944* (Oxford, Osprey, 2009)

Websites

http://www.beckettrankine.com
Marine Consulting Engineers

http://mulberrysurvey.co.uk
The Gooseberry, the newsletter of the archaeological survey of the artificial harbour built off Arromanches, Normandy

http://jproc.ca/hyperbolic/decca.html
Proc, Jerry (Ed), 'Hyperbolic radio navigation systems, Decca'

http://www.coppheroes.org
COPP Heroes of Hayling Island

http://www.ixengineercommand.com
US IX Engineer Command History

http://www.rquirk.com/cdnradar
Grande, George K., Linden, Sheila M., Macaulay, Horace R. (Eds), 'Canadians on Radar 1940–1945'

http://www.somerleyton.co.uk
'Duplex Drive amphibious tanks at Fritton Lake'

http://benbeck.co.uk/fh/transcripts/sjb_war_diaries/intro.html
Battery Diary, 341 Battery, 86th Field Regiment, Royal Artillery, 1944 to 1946

Useful contacts

D-DAY MUSEUM AND OVERLORD EMBROIDERY
Clarence Esplanade, Southsea, Hants PO5 3NT, UK
www.ddaymuseum.co.uk   023 9282 7261
Portsmouth's D-Day Museum is Britain's only museum dedicated solely to covering all aspects of the D-Day landings.

THE NATIONAL WWII MUSEUM
945 Magazine Street, New Orleans, LA 70130, USA
www.nationalww2museum.org   001 (504) 528-1944
Founded by historian and author, Stephen Ambrose, the museum tells the story of the American Experience in the Second World War and showcases large artefacts of the war, including exhibits about D-Day.

LE MÉMORIAL DE CAEN
Esplanade Général Eisenhower, BP 55026, 14050 Caen Cedex 4, FRANCE
http://www.memorial-caen.fr   +33(0)2 31 06 06 44
This is a museum for peace that gives an overall history of war from 1918 to the present day, including an exhibition dedicated to D-Day and the Battle of Normandy.

CINÉMA CIRCULAIRE ARROMANCHES 360
Chemin du Calvaire, BP 9, 14117 Arromanches, FRANCE
www.arromanches360.com   +33(0)2 31 22 30 30
The film *The Price of Liberty* that mixes wartime footage with views of the battlefield today can be seen in a 360° cinematic panorama using nine huge screens arranged in a circle.

JUNO BEACH CENTRE
Voie des Français Libres, BP 104, 14470 Courseulles-sur-Mer, FRANCE
www.junobeach.org   +33(0)2 31 37 32 17
The Juno Beach Centre presents the war effort made by all Canadians, both civilian and military, at home and on the various battle fronts during the Second World War.

Index